MATH SENSE

3 FOCUS ON ANALYSIS

Math for the High School Equivalency Tests

f Σ ∞ $\sqrt{}$

Cathy Fillmore Hoyt

New Readers Press
ProLiteracy's publishing division

Math Sense Book 3: Focus on Analysis
ISBN 978-1-56420-693-0

New Readers Press
ProLiteracy's Publishing Division
101 Wyoming Street, Syracuse, New York 13204
www.newreaderspress.com

Printed in the United States of America
10 9 8

Proceeds from the sale of New Readers Press materials support professional development, training, and technical assistance programs of ProLiteracy that benefit local literacy programs in the U.S. and around the globe.

Math Sense First Edition Developer: Learning Unlimited
Content Editor: Kathy Osmus
Editorial Director: Terrie Lipke
Illustrator: Kerry Manahan
Technology Specialist: Maryellen Casey

TI-30XS MultiView™ Scientific Calculator image courtesy of Texas Instruments

CONTENTS

INTRODUCTION

Math skills play an increasingly vital role in today's world. Everyone needs to work confidently with numbers to solve problems on the job and in daily life.

The increasing role of mathematics is reflected on the GED® and HiSET® mathematics tests. This book is part of a three-book series designed to help you pass these tests and prepare for new opportunities in education and employment.

In *Math Sense Book 3: Focus on Analysis*, you will acquire skills in areas such as data analysis, probability, systems of equations, functions, polynomials, and quadratics. Each unit is organized around four key areas that will build your competence and confidence.

- **Skills** pages present instruction and practice with both computation and word problems.
- **Tools** pages provide insight on how to use calculators, key ideas, and mathematical properties to solve math problems.
- **Problem Solver** pages present strategies that will help you find the best way to approach different types of problems.
- **Test Taker** pages provide math tips and problem-solving approaches that good test takers use to take standardized tests.

KEY FEATURES

SKILL PREVIEW The Skill Preview can help you determine what skills you already have and identify which areas you most need to concentrate on.

UNIT PREVIEW The beginning of each unit features questions and topics to write about or discuss with classmates. Talking about math is key to building your understanding.

CORE CONNECTIONS Throughout the book, you will work with topics that connect math ideas to real life and to other math concepts. These Core Connections add depth to your learning and correspond to the national Common Core State Standards (CCSS).

SPECIAL PROBLEMS These specially labeled problems require an in-depth exploration of math ideas. You may be asked to explain your reasoning or to find multiple solutions. These problems will prepare you for the short-response items on the GED Mathematics Test.

MIXED REVIEWS AND UNIT REVIEWS Periodic checkups will help you see how well you understand and can apply the material. Unit Reviews also allow you to practice the types of questions that you will see on the GED Mathematics Test, including:

- Multiple-choice questions
- Fill-in-the-blank questions
- Drop-down and matching questions
- Short-response questions

GED SIMULATED TEST At the end of the book is a 25-item simulated GED Mathematics Test that focuses on the topics from this book only. You can use this final assessment to judge how well you have mastered the skills and strategies presented in this book.

If you are taking the HiSET test, go to newreaderspress.com. On the *Math Sense* product page, you can download a free practice test.

GLOSSARY This list of terms defines key math words and ideas.

TOOL KIT These resource pages provide helpful information you can use as you work through the book.

Skill Preview

This survey of math skills will help you and your teacher decide what you need to study to get the most out of this book. It will show you how much you already know and what you need to learn.

Do as much as you can of each section. If you can't do all of the problems in a section, go ahead to the next section and do all of the problems that you can.

PART 1: DATA ANALYSIS

Use the bar graph for problems 1 and 2.

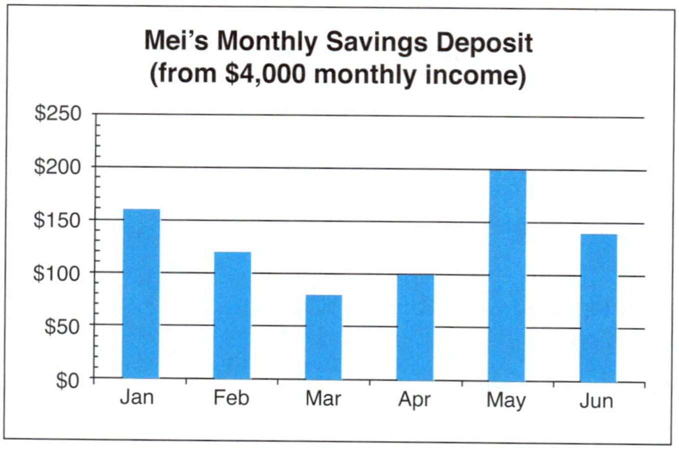

1. How many more dollars did Mei put in savings in May than she did in April?

2. What percent of her monthly income did Mei put in savings in January?

Use the circle graph for problems 3 and 4.

Pizza Bob offers five kinds of pizza. The circle graph shows the percent of each type of pizza ordered.

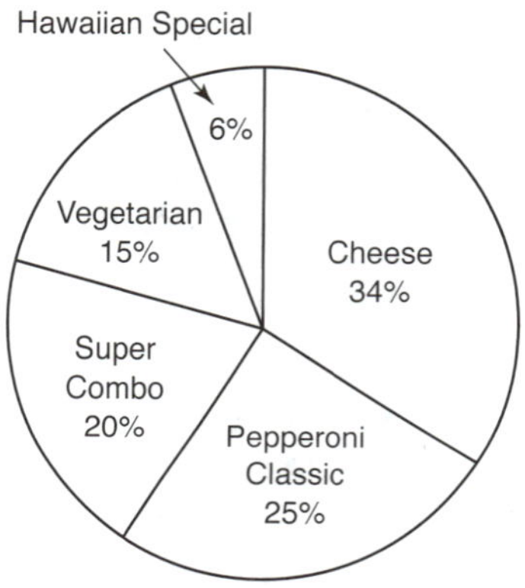

Percent of Pizzas Ordered by Type

3. If 1,500 pizzas were ordered last month, how many were cheese pizzas?

4. The three most popular pizzas combine to make up what percent of the orders?

Use the table for problem 5.

Lucky Beans Coffee Sales
Week of August 4

	Espresso	Green Tea
Monday	29	16
Tuesday	25	25
Wednesday	33	21
Thursday	41	32
Friday	30	19

5. In all, how many more espressos than green teas were sold on Wednesday, Thursday, and Friday?

Use the following information for problems 6–8.

Marissa is buying a backpack to take on a trip. The ten leading brands have the following prices: $23, $29, $19, $33, $39, $29, $55, $29, $35, and $39.

6. Find the mean price.

7. What is the median price of the data set?

8. What is the mode of the prices?

Use the line graph for problems 9 and 10.

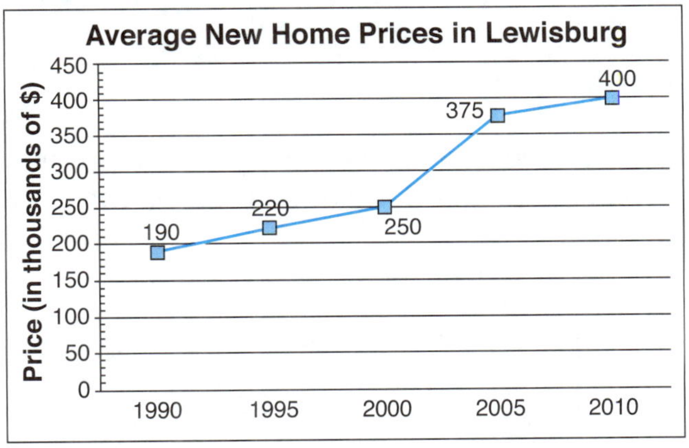

9. Estimate the average new home price in 2003.

10. Analysts predict that the price change from 2010 to 2020 will be similar to the price change from 1990 to 2000. Which is the best estimate for the average new home price in 2020?

A. $250,000
B. $435,000
C. $460,000
D. $550,000

Use the box plot for problems 11 and 12.

Over the weekend, Sports Depot held a shoe sale. The box plot shows the distribution of men's sizes sold over the weekend.

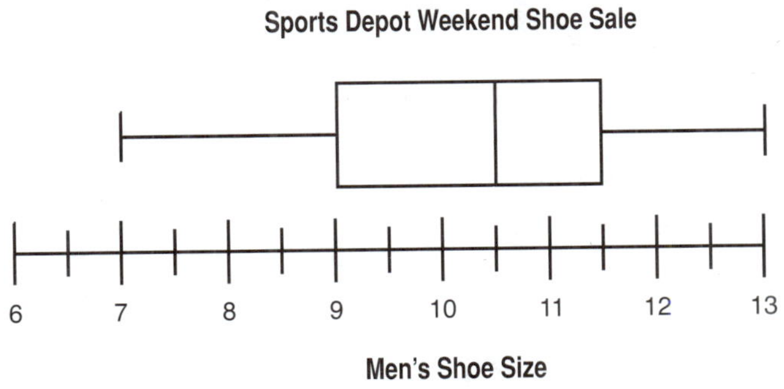

11. What was the median men's shoe size sold over the weekend?

12. If 200 pairs of men's shoes were sold during the sale, about how many were from size 7 to size 9?

PART 2: COUNTING AND PROBABILITY

Solve as directed.

13. Jenna is designing a flag using three different colors: A, B, and C. Color C is used in two regions.

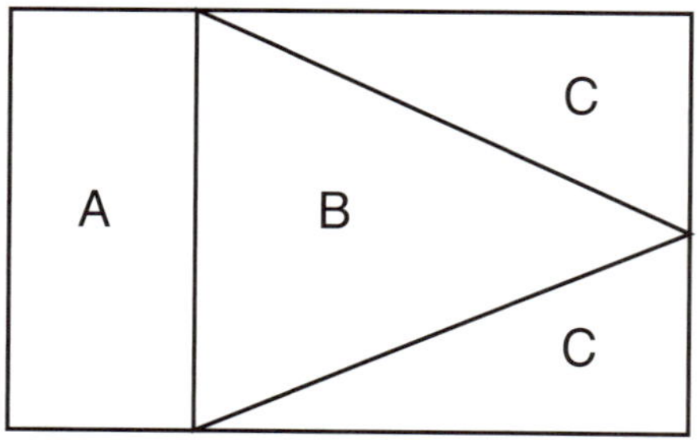

She can choose from these colors: white, green, black, red, blue, and yellow. How many different flags can she create?

14. Of the 12 players on a volleyball team, 2 will be chosen as co-captains. How many different pairs of co-captains can be selected?

15. At the football combine, 5 athletes compete in the 40-yard dash. How many different ways can the runners finish the race?

Use the spinner for problems 16 and 17.

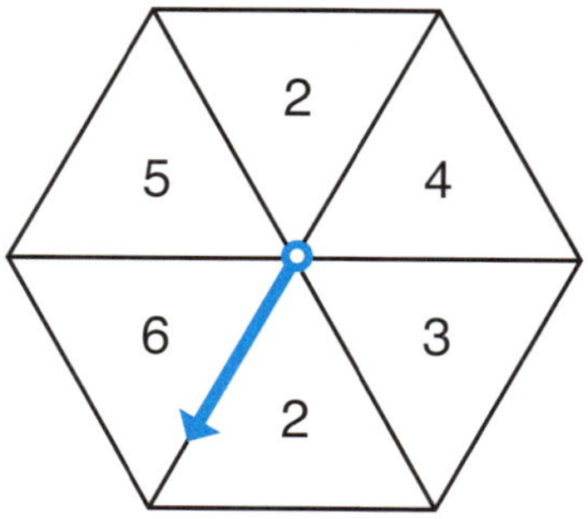

16. The spinner is divided into 6 equal sections. If you spin the spinner once, what is the probability of getting an even number?

17. If you spin the spinner twice, what is the probability that you will get 2 both times?

18. Jim has to take a quiz with 5 true-false questions. He didn't study and decides to guess randomly on each question. What is the percent chance that he won't get any questions right?

PART 3: SYSTEMS OF EQUATIONS AND INEQUALITIES

Solve each system for *x* and *y* in problems 19–22.

19. $\begin{cases} y = 4x - 5 \\ 6x + 5y = 1 \end{cases}$

20. $\begin{cases} -19 + y = 10x \\ 4x + 10 = -2y \end{cases}$

21. $\begin{cases} -6x - 3y = 15 \\ -4x - 9y = -11 \end{cases}$

22. $\begin{cases} 7x - 8y = -24 \\ x + 2y = -16 \end{cases}$

Solve.

23. In a right triangle, angle x is 26° greater than angle y. The sum of the angles in a triangle is 180°. What is the measure of angle x?

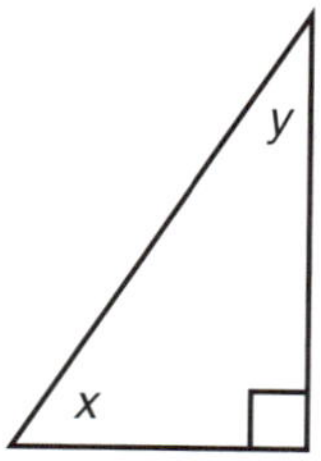

24. A hotel has 160 rooms. Rooms with 2 beds are $90 per night. Deluxe suites are $135 per night. During a conference, all the rooms are filled. If the hotel receives $16,425 for 1 night, how many deluxe suites does the hotel have?

25. One linear equation from a system is graphed on the coordinate plane below.

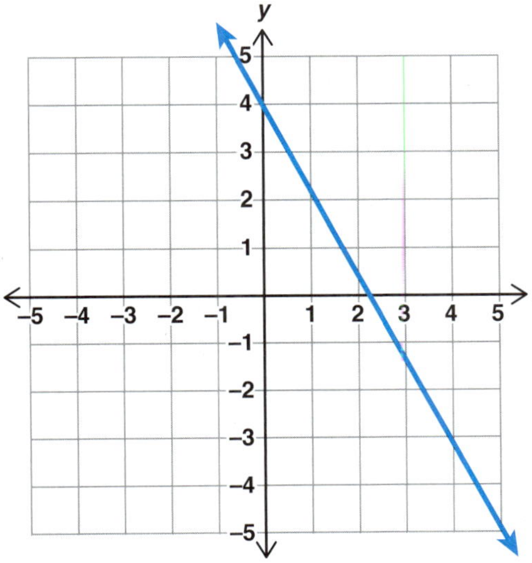

If the other equation in the system is $x = 4$, what is the solution to the equation? Write the solution as an ordered pair (x, y).

26. Which of the following systems is graphed on the coordinate plane below?

A. $\begin{cases} y < 2x - 3 \\ y \geq -3x + 2 \end{cases}$

B. $\begin{cases} y \geq 3x - 2 \\ y \leq 2x - 3 \end{cases}$

C. $\begin{cases} y > -3x + 2 \\ y \leq 2x - 3 \end{cases}$

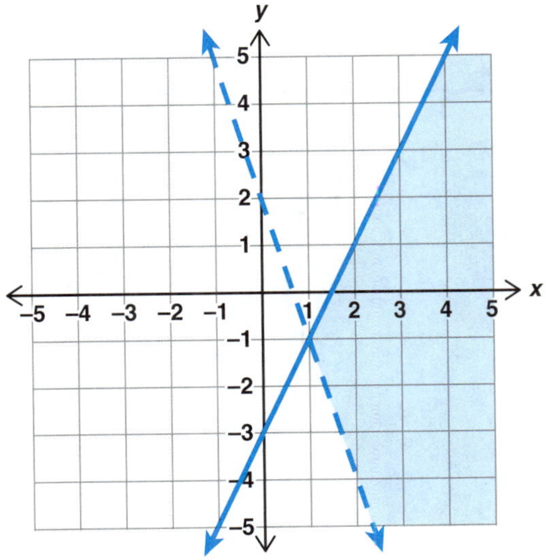

27. Shade the region that represents the solution set to the following system.

$\begin{cases} x - 2y \leq 4 \\ 2x - y \geq -1 \end{cases}$

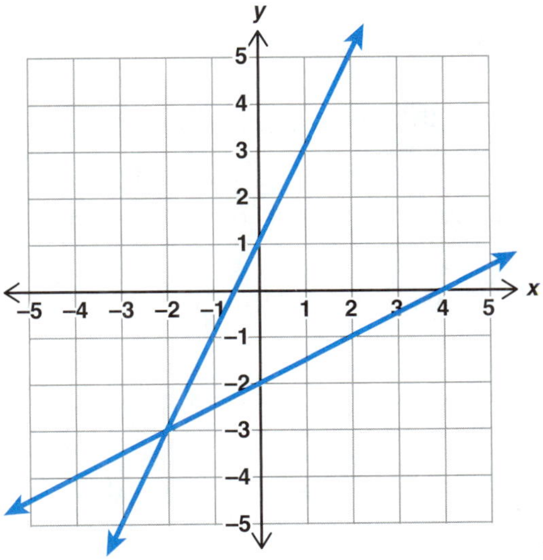

28. Using the graph below, write an inequality symbol in each box to complete the system.

Choose from these symbols: >, ≥, <, and ≤.

$\begin{cases} x + 3y \ \square \ 6 \\ x - y \ \square \ 2 \end{cases}$

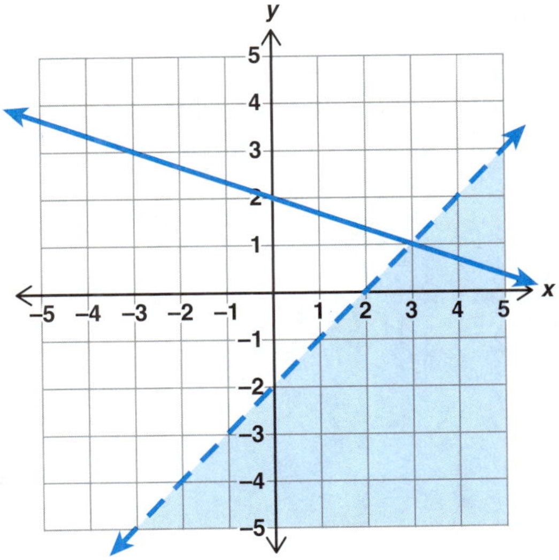

PART 4: FUNCTIONS

Solve as directed.

29. If the pattern below continues, what will be the 24th symbol in the sequence?

A B 1 A B 2 A B 3 A . . .

30. Find the 17th term in the sequence. $-2, -6, -10, -14, \ldots$

31. What is the next term in the sequence? $2, -6, 18, -54, 162, \ldots$

32. Find $f(11)$ when $f(n) = 2n - 5$.

33. Find $g(-2)$ when $g(x) = x^3 - x^2$.

34. Find $S(0)$ when $S(t) = 3t^2 + 11t - 19$.

35. In baseball, a player is awarded a putout (P) when he gets a player out and an assist (A) when he makes a throw that gets a player out. A player's Range Factor (RF) is found using this formula: $RF = \frac{P + A}{G}$, where G is the number of games played. A third baseman played in 67 games and had 35 putouts and 151 assists. To the nearest hundredth, what is his Range Factor?

Use the graph for problems 36–38.

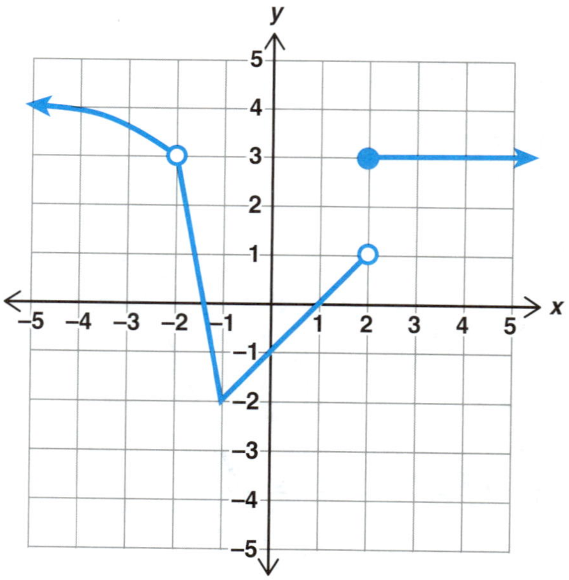

36. The graph shows the function h. What is the value of $h(0)$?

37. For which value of x does the function have no value?

38. Which interval on the graph shows the least amount of change?

A. $x < -2$

B. $-2 < x < -1$

C. $-1 < x < 2$

D. $x \geq 2$

PART 5: POLYNOMIALS AND RATIONAL EXPRESSIONS

39. Simplify the expression.

$(3b^2 + 4) - (4 - 3b^2) - (7b^2 + 7)$

40. Multiply.

$-7x(3x^2 - 4x + 2)$

41. Factor the expression.

$n^2 + n - 20$

42. Factor the expression.

$2x^2 + 15x - 27$

43. Add.

$\frac{2}{n} + \frac{n-6}{5n+3}$

44. Simplify by factoring.

$\frac{x^2 + 4x - 21}{x - 3}$

45. What numbers must be excluded from the solution set of this expression?

$\frac{n+3}{2n}$

46. Simplify the expression and state which numbers must be excluded from the solution set.

$\frac{x-5}{x^2 + 5x - 50}$

PART 6: QUADRATIC EQUATIONS

Find all solutions to the equations in problems 47–49.

47. $4x^2 = 36$

48. $n^2 - 2n - 48 = 0$

49. $3p^2 + 12 = -20p$

50. The equation $d^2 = \frac{3}{2}h$ can be used to estimate how far you can see as you look out over the ocean. The distance d (in miles) that you can see depends on the height h (in feet) of your eyes above sea level. If your eyes are 24 feet above sea level, for how many miles can you see?

Answers start on page 12.

Skill Preview Answer Key

1. **$100**
$\$200 - \$100 = \$100$

2. **4%**
$\frac{\$160}{\$4{,}000} = 4\%$

3. **510 cheese pizzas**
34% of 1,500 = $1{,}500 \times 0.34 = 510$

4. **79%**
$34\% + 25\% + 20\% = 79\%$

5. **32 more espressos**
Espressos: $33 + 41 + 30 = 104$
Green teas: $21 + 32 + 19 = 72$
$104 - 72 = 32$

6. **$33**
$\frac{\$23 + \$29 + \$19 + \$33 + \$39 + \$29 + \$55 + \$29 + \$35 + \$39}{10} = \$33$

7. **$31**
Put the values in order:
$19, $23, $29, $29, $29, $33, $35, $39, $39, $55
The two middle values are $29 and $33.
$\frac{\$29 + \$33}{2} = \$31$

8. **$29**

9. Answers should be between $300,000 and $340,000.

10. **C. $460,000**
1990 to 2000: $\$250{,}000 - \$190{,}000 = \$60{,}000$
2010 to 2020: $\$400{,}000 + \$60{,}000 = \$460{,}000$

11. **10.5**
On a box plot, the median is the center line of the box. The center line of this box plot is at 10.5.

12. **50**
25% of the data in a box plot falls between the minimum and the lower quartile. 25% of 200 is 50.

13. **120 flags**
Order matters. There are only 3 choices for colors even though there are 4 sections on the flag.
$6 \times 5 \times 4 = 120$

14. **66 pairs**
Order doesn't matter, so find the number of combinations.
$\frac{12 \times 11}{2 \times 1} = 66$

15. **120 ways**
Order matters. Find the number of permutations.
$5 \times 4 \times 3 \times 2 \times 1 = 120$

16. $\frac{4}{6} = \mathbf{\frac{2}{3}}$
4 out of the 6 sections have an even number.

17. $\mathbf{\frac{1}{9}}$
The probability of spinning one 2 is $\frac{2}{6} = \frac{1}{3}$.
$\frac{1}{3} \times \frac{1}{3} = \frac{1}{9}$

18. **3.125%**
Jim has a 1 in 2 chance of missing a question.
$\frac{1}{2} \times \frac{1}{2} \times \frac{1}{2} \times \frac{1}{2} \times \frac{1}{2} = \frac{1}{32} = 3.125\%$

19. **$x = 1$, $y = -1$**
Substitute for y in the second equation.
$$6x + 5(4x - 5) = 1 \qquad y = 4(1) - 5$$
$$6x + 20x - 25 = 1 \qquad y = -1$$
$$26x = 26$$
$$x = 1$$

20. **$x = -2$, $y = -1$**
Solve for y in the first equation: $y = 10x + 19$.
Substitute for y in the second equation.
$$4x + 10 = -2(10x + 19) \qquad -19 + y = 10(-2)$$
$$4x + 10 = -20x - 38 \qquad y = -20 + 19$$
$$24x = -48 \qquad y = -1$$
$$x = -2$$

21. **$x = -4$, $y = 3$**
Multiply the first equation by –3 and eliminate.
$$\begin{array}{r} 18x + 9y = -45 \\ -4x - 9y = -11 \\ \hline 14x \quad = -56 \\ x = -4 \end{array} \qquad \begin{array}{r} -4(-4) - 9y = -11 \\ 16 - 9y = -11 \\ -9y = -27 \\ y = 3 \end{array}$$

22. $x = -8, y = -4$

Multiply the second equation by 4 and eliminate.

$$\begin{array}{r} 7x - 8y = -24 \\ 4x + 8y = -64 \\ \hline 11x \quad = -88 \\ x = -8 \end{array} \qquad \begin{array}{r} 7(-8) - 8y = -24 \\ -56 - 8y = -24 \\ -8y = 32 \\ y = -4 \end{array}$$

23. $x = 58°$

Write a system of equations.

$$\begin{cases} x = 26 + y \\ x + y + 90 = 180 \end{cases}$$

Substitute for x in the second equation.

$$\begin{array}{r} 26 + y + y + 90 = 180 \\ 2y + 116 = 180 \\ 2y = 64 \\ y = 32 \end{array} \qquad \begin{array}{l} x = 32 + 26 \\ x = 58 \end{array}$$

24. 45 deluxe suites

Write a system of equations. Let x = rooms with 2 beds, and let y = deluxe suites.

$$\begin{cases} x + y = 160 \\ 90x + 135y = 16{,}425 \end{cases}$$

Multiply the first equation by –90.

$$\begin{array}{r} -90x - 90y = -14{,}400 \\ 90x + 135y = 16{,}425 \\ \hline 45y = 2{,}025 \\ y = 45 \end{array}$$

25. (4, –3)

On the graphed line, when $x = 4$, $y = -3$. The point of intersection would by (4, –3).

26. C. $\begin{cases} y > -3x + 2 \\ y \le 2x - 3 \end{cases}$

The shaded portion is less than the solid line, so this represents $y \le 2x - 3$.

The shaded portion is greater than the dashed line, so this represents $y > -3x + 2$.

27.

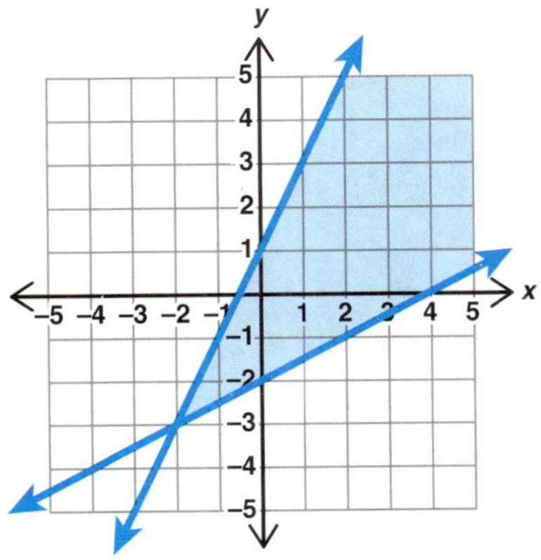

28. ≤, >

The shaded portion is less than the solid line, so this represents $y \le -\frac{1}{3}x + 2$, or $x + 3y \le 6$.

The shaded portion is less than the dashed line, so this represents $y < x - 2$, or $x - y > 2$.

29. 8

Every third term in the pattern is a number. Since 24 is a multiple of 3, the 24th term will be a number. $24 \div 3 = 8$, so the 24th term will be 8.

30. –66

This sequence is made by adding –4. You would need to add –4 sixteen times to get the 17th term.

$-4 \times 16 = -64$

$-2 + -64 = -66$

31. –486

Each term is multiplied by –3.

$162 \cdot -3 = -486$

32. 17

$f(11) = 2(11) - 5$

$f(11) = 17$

33. –12

$g(-2) = (-2)^3 - (-2)^2$

$g(-2) = -12$

34. –19

$S(0) = 3(0)^2 + 11(0) - 19$

$S(0) = -19$

35. 2.78

$RF = \frac{35 + 151}{67} \approx 2.78$

36. –1

37. –2

38. D. $x \ge 2$

When $x \ge 2$, the function is constant at 3.

39. $-b^2 - 7$

$(3b^2 + 4) - (4 - 3b^2) - (7b^2 + 7)$

$= 3b^2 + 4 - 4 + 3b^2 - 7b^2 - 7$

$= -b^2 - 7$

40. $-21x^3 + 28x^2 - 14x$

41. $(n + 5)(n - 4)$

42. $(2x - 3)(x + 9)$

43. $\frac{n^2+4n+6}{n(5n+3)}$ **or** $\frac{n^2+4n+6}{5n^2+3n}$

$\frac{2(5n+3)+n(n-6)}{n(5n+3)} = \frac{10n+6+n^2-6n}{n(5n+3)} = \frac{n^2+4n+6}{n(5n+3)}$

44. $\frac{(x+7)\cancel{(x-3)}}{\cancel{x-3}}$ $= \mathbf{x + 7}$

45. 0

The denominator of the expression equals 0 when $n = 0$.

46. $\frac{1}{x+10}$**, exclude** $x = -10$, $x = 5$

Factor: $\frac{x-5}{(x+10)(x-5)}$

The denominator of the expression equals 0 when $x = -10$ and $x = 5$.

Simplify: $\frac{\cancel{x-5}}{(x+10)\cancel{(x-5)}} = \frac{1}{x+10}$

47. 3, –3

$4x^2 = 36$

$x^2 = 9$

$x = \pm 3$

48. 8, –6

$(n-8)(n+6) = 0$

$n-8=0 \qquad n+6=0$

$n = 8 \qquad n = -6$

49. $-\frac{2}{3}$**, –6**

$3p^2 + 12 = -20p$

$3p^2 + 20p + 12 = 0$

$(3p+2)(p+6) = 0$

$3p+2=0 \qquad p+6=0$

$p = -\frac{2}{3} \qquad p = -6$

50. 6 miles

$d^2 = \frac{3}{2} \cdot 24$

$d^2 = 36$

$d = 6$

Skill Preview Diagnostic Chart

Make note of any problems that you answered incorrectly. As you work through this book, be sure to focus on these skill areas.

Problem Number	Skill Area	Unit
1, 2, 3, 4, 5	Working with Graphs and Charts	1
6, 7, 8	Mean, Median, and Mode	1
9, 10, 11, 12	Drawing Conclusions from Data	1
13, 14, 15	Permutations and Combinations	2
16, 17, 18	Probability	2
19, 20, 21, 22, 25	Systems of Equations	3
23, 24	Solving Word Problems with Systems	3
26, 27, 28	Linear Inequalities	3
29, 30, 31	Patterns and Sequences	4
32, 33, 34, 35	Functions	4
36, 37, 38	Working with Graphs of Functions	4
39, 40, 41, 42	Working with Polynomials	5
43, 44, 45, 46	Working with Rational Expressions	5
47, 48, 49, 50	Solving Quadratic Equations	6

UNIT 1

DATA ANALYSIS

SKILLS

- Reading tables and graphs
- Reading histograms and scatter plots
- Finding mean, median, and mode
- Calculating weighted averages
- Reading box plots
- Analyzing distribution
- Seeing trends and making predictions
- Analyzing correlation
- Identifying a representative sample
- Using more than one data source

TOOLS AND STRATEGIES

- Applying typical values
- Drawing conclusions from data
- Analyzing misleading graphs

THE TERM *DATA* describes a collection of facts. When the facts are numerical, you can use mathematics to analyze and draw conclusions from the data.

Data is easier to analyze when it is organized. A **graph** is a visual representation of data that organizes the information and provides a quick overview of the main points. In this unit, you'll study several of the most commonly used types of graphs.

Data can also be easier to analyze when it is summarized. **Typical values** represent what is normal, or average, in a set of data. You'll learn three types of typical values and how to use them to draw conclusions.

You can also use patterns in the data to draw conclusions. In this unit, you'll learn to recognize patterns and make predictions based on those patterns.

USING DATA ANALYSIS

You use data analysis when you:

- Read a table to find out how much medicine to give a child
- Analyze sports statistics to pick the winner of a game or event
- Interpret a graph in a magazine or newspaper
- Decide whether to believe the information in a sales pitch
- Read a bus or train schedule

Read the following items. Describe some of your experiences in writing or discuss them with a partner.

1. Look through a newspaper or a magazine and count how many graphs you find. Which graph is easiest to read? Why?

2. Have you ever calculated your own grade in a class or your grade point average (GPA)? How do you determine what your average should be?

3. Think of some things that increase when the temperature increases. For example, when the temperature increases, the number of people at a water park increases. What things decrease when it gets hot out?

4. Do you trust the graphs you see in newspapers or magazines or on the Internet? What makes a graph trustworthy? What makes it misleading?

MATH TALK

In real life, people are not usually interested in an entire set of data. They want to know what's normal, usual, or typical. For example, if someone asks you about how much a large pizza costs, they probably don't want to know the price of a large pizza at every restaurant in the city. They just want to know that a large pizza normally costs about $15. They want to know the typical value.

With a partner, make a list of situations where typical values are all you really need to know. Then make a list of situations where a typical value isn't enough information.

WORKING WITH TABLES

A **table** is often used to organize and display data, especially data that is classified by categories. A table may contain both words and numbers written in labeled rows and columns.

A **row** is read from left to right. A **column** is read from top to bottom.

Use the row and column headings to locate specific data in a table. The title of the table may also have important information.

Population of the Top 5 U.S. Cities (2000 and 2010)
(in millions)

City	2000 Population	2010 Population	2010 Rank	Percent Change 2000–2010
Chicago, IL	2.89	2.69	3	–6.9%
Houston, TX	1.95	2.10	4	+7.5%
Los Angeles, CA	3.69	3.79	2	+2.6%
New York, NY	8.01	8.17	1	+2.1%
Philadelphia, PA	1.51	1.52	5	+0.6%

Source: U.S. Census, 2000 and 2010

FINDING INFORMATION IN TABLES

Example In 2000, the population of New York City was about how many times larger than the population of Houston?

Locate the place where the column labeled “2000 Population” and the row labeled “New York, NY” intersect.

	2000 Population
	2.89
	1.95
	3.69
New York, NY	8.01
	1.51

The value is 8.01. The title states that the values are in millions. The population of New York City was about 8.01 million in 2000.

Now look for the row labeled “Houston, TX.” The population of Houston was about 1.95 million in 2000.

$8.01 \div 1.95 \approx 8 \div 2 \approx 4$. The population of New York City was about **4 times** larger than the population of Houston in 2000.

You can also use the data in a table to draw conclusions. For example, look at the column labeled “Percent Change.” Chicago was the only city with a decreasing population. You could conclude that from 2000 to 2010, the populations of the largest U.S. cities primarily increased.

Use the following table for problems 1 and 2.

Summary of Saturday Price Reductions

Item	Original Price	Discount Rate
Portable Speaker	$80	15%
Earbuds	$18	5%
MP3 Player	$26	10%
DVD Player	$120	25%
HD Television	$230	20%

1. What is the difference in price between a portable speaker and a DVD player on Saturday?

2. Is the actual discount (in dollars) on an HD television greater than or less than the actual discount on a DVD player?

Use the table for problems 3–6.

Yard Waste Recycled by County (in tons)

Month	Carroll County	Logan County
January	520	1,345
February	1,080	2,100
March	990	2,980
April	2,250	3,405

3. How much more yard waste did Logan County recycle in February than Carroll County?

4. During which month did Carroll County's recycling decrease compared to the previous month?

5. During which month did Logan County's recycling increase the most compared to the previous month?

6. From January to April, which county's recycling increased the most, and by how much?

Use the mileage chart for problems 7 and 8.

Airline Mileage to North American Cities

Cities	Mexico City	Montreal	New York	San Francisco
Mexico City		2312	2085	1883
Montreal	2312		330	2537
New York	2085	330		2566
San Francisco	1883	2537	2566	

7. How many miles will Armando travel if he flies from Mexico City to San Francisco and continues to New York?

8. **Explain** Why are there four empty boxes on this chart?

Answers start on page 163.

BAR GRAPHS AND HISTOGRAMS

Graphs give a visual picture of data. In a **bar graph**, data is represented by the height (or length) of **data bars**. The bars may be drawn vertically or horizontally.

For the bar graph below, you can use the **title** (Cafeteria Orders at Newark High School) and axis labels to determine what values are being compared. The **horizontal axis** shows the type of meal ordered. The **vertical axis** shows the number of meals ordered during the week.

FINDING INFORMATION ON BAR GRAPHS

Example About how many orders of chicken finger meals were sold at Newark High School from 9/15 to 9/19?

Find the bar labeled Chicken Fingers. Read the value on the vertical axis closest to the height of the bar.

You may need to estimate values on a bar graph. The bar for chicken fingers is about halfway between 240 and 250. **About 245 chicken finger meals** were sold from 9/15 to 9/19.

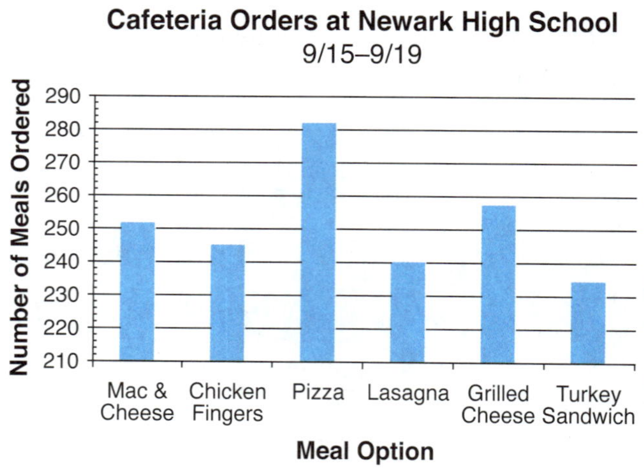

Use the bar graph above for problems 1 and 2.

1. Which meal options sold over 250 meals from 9/15 to 9/19?

2. About how many more orders were for pizza than for lasagna?

Use the double bar graph for problems 3–6.

3. About how much more did the petting zoo make in Year 2 than Year 1?

4. In which event did profits decrease from Year 1 to Year 2?

5. In Year 1, about how much more money did the auction make than the food booths?

6. Which event's profits increased the most from Year 1 to Year 2?

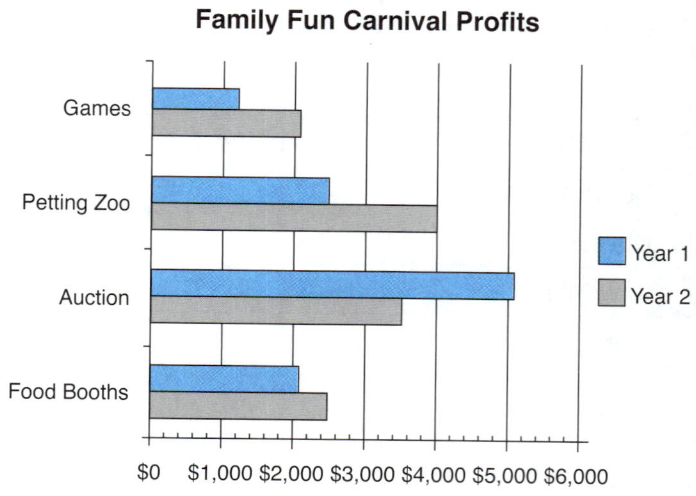

A **histogram** also uses data bars to visually show data. On a bar graph, the bars represent categories. On a histogram, the bars represent a range of numbers. You can use a histogram to draw conclusions about how frequently events occur.

Each bar on the histogram below represents an interval. For example, the first bar represents the number of cars sold with an estimated gas mileage of 5 to 9 miles per gallon. Use the vertical axis to read the value of the first bar. Two cars were sold with an estimated gas mileage between 5 and 9 miles per gallon.

FINDING INFORMATION ON HISTOGRAMS

Example How many cars were sold in April with an estimated gas mileage of less than 15 miles per gallon?

Find 15 on the horizontal axis. The bars to the left of this marker represent cars with a mileage of less than 15 mpg.

Read the values for both bars and find the total.

5 to 9: 2
10 to 14: 4
Total: 2 + 4 = 6

In April, **6 cars** with an estimated gas mileage of less than 15 mpg were sold.

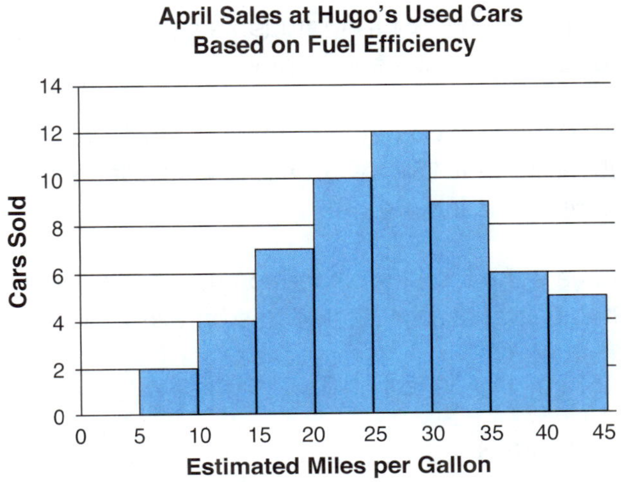

Use the histogram above for problems 7 and 8.

7. Did Hugo's Used Cars sell more cars with an estimated mileage greater than 25 mpg or less than 25 mpg?

8. How many cars did Hugo's Used Cars sell in April?

Use this histogram for problems 9 and 10.

9. A grade below 70% is a failing grade. How many students failed the final?

10. **Explain** Students who scored above 70% passed the final. To the nearest percent, what percent of Ms. Jung's class passed the final? Explain your answer.

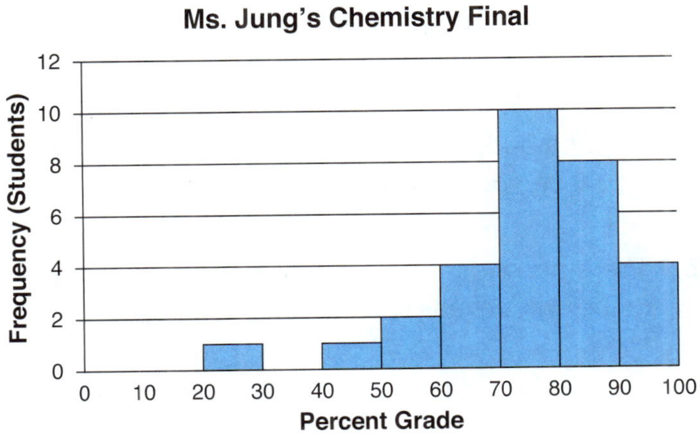

Answers start on page 163.

CIRCLE GRAPHS

A **circle graph** displays data as sections of a circle. Circle graphs are especially useful to show how a whole amount, such as a budget, is made up of several parts.

In a circle graph, the circle represents the whole amount. Each segment represents a part of the whole. Larger segments represent greater parts of the whole. Data in a circle graph can be displayed as a percent, a fraction, or a decimal. The whole circle can equal 100%, 1, or $1.00.

FINDING INFORMATION ON A CIRCLE GRAPH

Example How many students are enrolled as juniors at Grandview Community College?

Find the section of the graph labeled "Juniors." It shows that 15% of Grandview's students are enrolled as juniors.

Locate the total enrollment on the graph. Grandview has a total of 11,400 students enrolled during the fall.

15% of 11,400 is 1,710.

1,710 students are enrolled as juniors.

Fall Enrollment at Grandview Community College
Total Enrollment: 11,400 Students

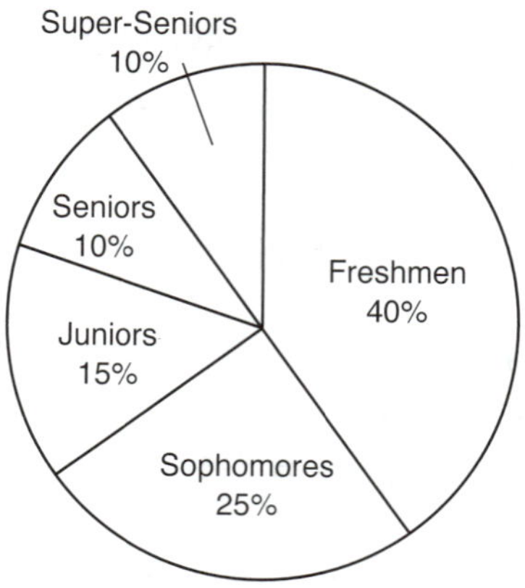

Use the circle graph above for problems 1 and 2.

1. How many students at Grandview Community College are either seniors or super-seniors?

2. Is the combined number of juniors, seniors, and super-seniors greater than or less than the number of freshmen?

Use this circle graph for problems 3–6.

3. Out of each budget dollar, how much did Central High School spend on track?

4. Out of a budget of $20,000, how much money would be spent on softball?

5. The amount spent on football equipment is equal to the amount spent on which three sports?

6. What percent of the equipment budget was spent on baseball?

Approximate Equipment Spending Central High School (cents per budget dollar)

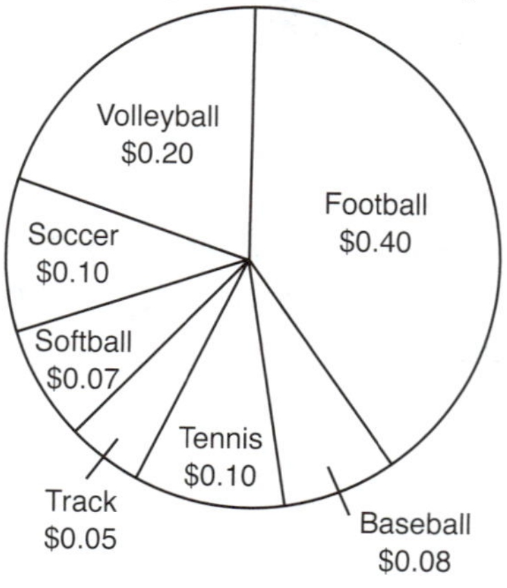

Use the bar graph and circle graph below to answer problems 7 and 8.

The graphs below both show Beverly's yearly car expenses.

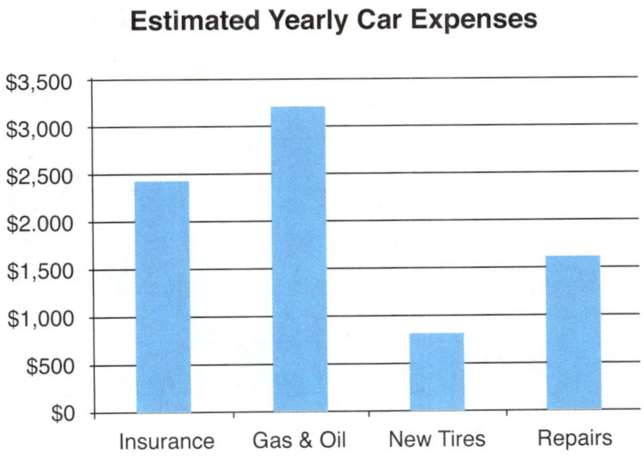

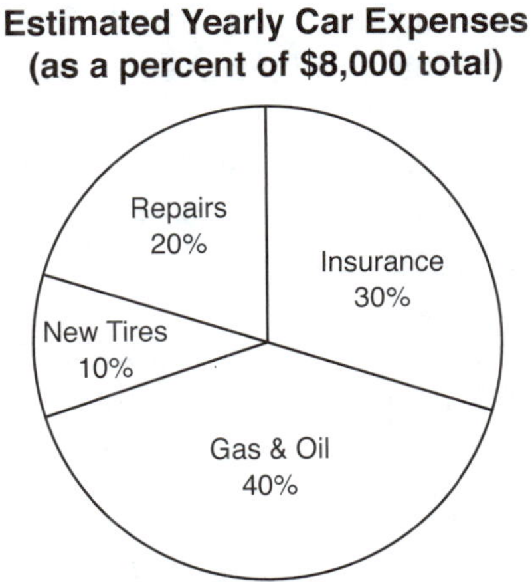

7. Name two expenses that combine to equal about one-half of Beverly's yearly budget. Which graph did you use?

8. **Explain** In which situations is a circle graph most useful? In which situations is a bar graph most useful?

CORE CONNECTIONS: Creating a Circle Graph

In a circle graph, the size of each segment depends on what percent of the total it represents. For example, a segment representing 20% of the whole will be 20% of the entire circle. To create a circle graph, find the number of degrees for each segment.

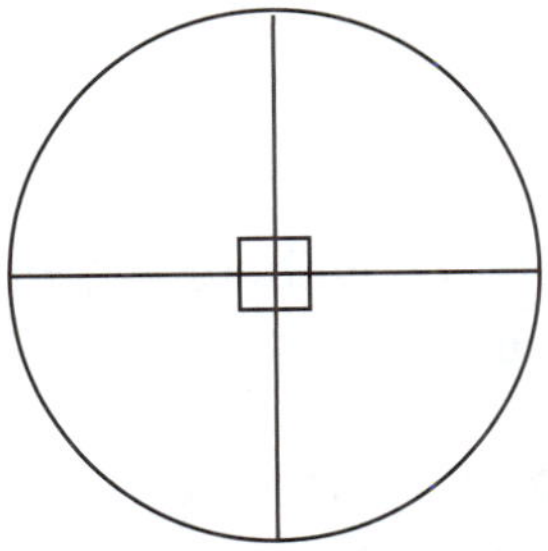

When a circle is divided into fourths, there are four right angles. You know that a right angle measures 90°. Since the total number of degrees in a circle equals the sum of all four angles, there are 360° in a circle: 4(90°) = 360°

You can find the size of a segment in a circle graph by taking a percent of 360°. Look at the graph of Estimated Yearly Car Expenses. The segment for car repairs is 20%. Since 20% of 360° is 72°, the size of the segment labeled "Repairs" measures 72°.

Complete the table to the right.

A total of 11,000 votes were cast in a recent city election. Use the total to find the percentage of votes for each candidate. Then use the percent to find the number of degrees it would take for each candidate's segment in a circle graph.

	Votes	Percent	Degrees
Kohli	3,300		
Sanderson	550		
Morales	4,950		
Souba	2,200		

Answers start on page 163.

SCATTER PLOTS

A **scatter plot** is a graph that shows the possible relationship between two different sets of data. Scatter plots are useful in graphing the data from a survey or poll.

The table shows the partial results of a survey. A total of twenty engineers were asked their years of experience in their field and their salary.

The results of the survey are represented on the scatter plot below. In this graph, the horizontal axis represents years of experience. The vertical axis represents salary.

Years of Experience	Salary
0	$55,000
2	$60,000
5	$70,000
10	$100,000
3	$60,000
8	$80,000
6	$75,000
5	$73,000

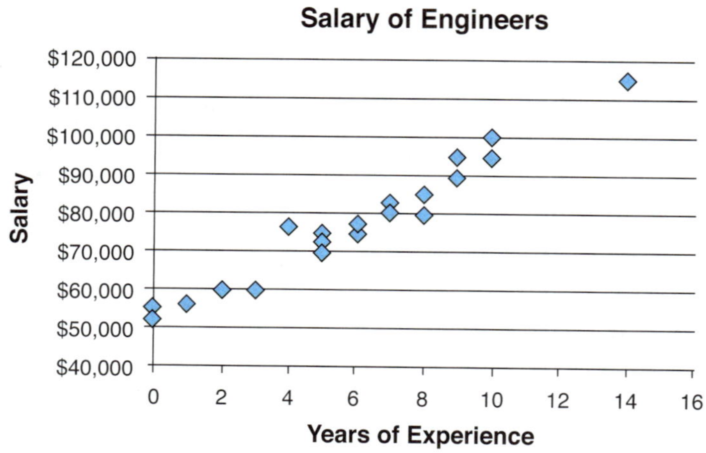

You can find individual points from the survey on the graph, but a scatter plot is meant to show the relationship between the data sets.

Is there a relationship between an engineer's experience and salary? The data on the graph show that the more experience an engineer has, the higher the salary.

You can say there is a **correlation** between years of experience and salary. You could even identify the type of correlation.

A **positive correlation** means that when values in one set of data increase, the values in the other set also increase.

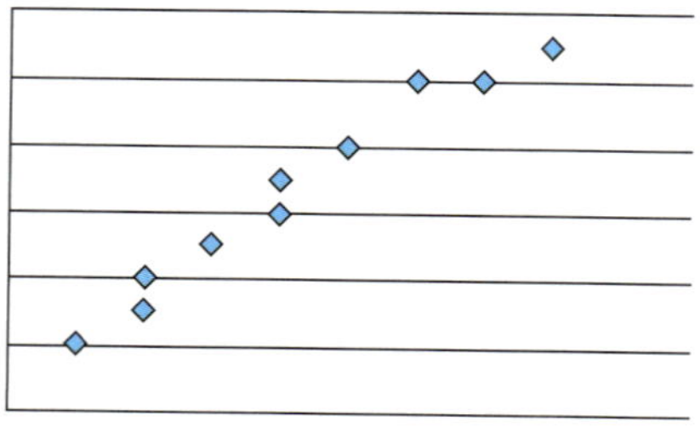

Years of experience and an engineer's salary have a positive correlation.

A **negative correlation** means that when values in one set of data increase, the values in the other set of data decrease.

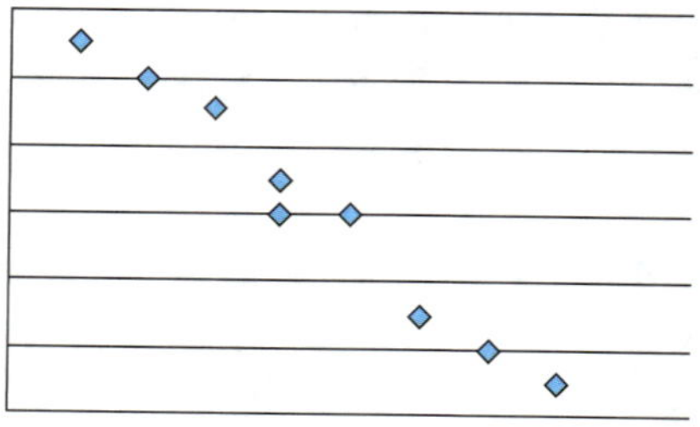

The number of days absent and a student's grades might have a negative correlation.

Sometimes sets of data have no relationship, or **no correlation**.

Hat size and the number of pets owned have no correlation.

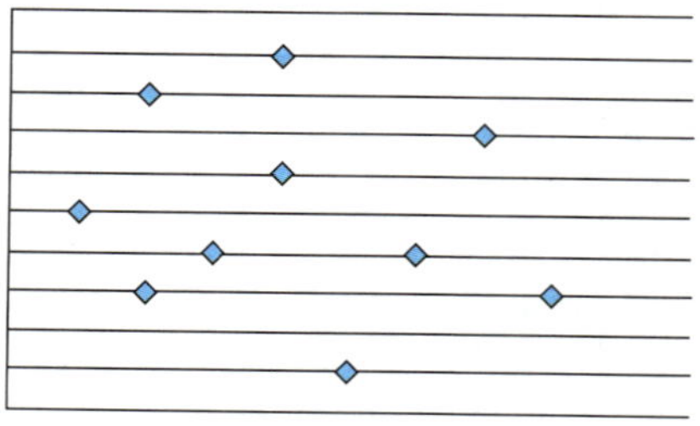

Use the scatter plot for problems 1 and 2.

The scatter plot shows the results of a study comparing the maximum distance from which a driver can read a street sign and the age of the driver.

1. Is there a correlation between the two groups of data? If so, is it positive or negative?

2. Theo is 35 years old. Based on the results of the study, predict the maximum distance from which he can read a street sign.

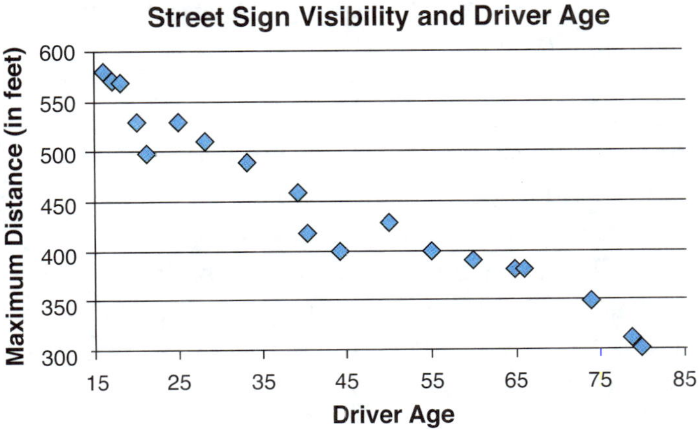

Use the table and scatter plot below for problems 3 and 4.

A company requires its employees to take a policy test every two years. Employees can study an online handbook that tracks hours. The results are listed in the table below.

3. Complete the scatter plot using the data in the table.

Hours of Study	Test Score
24	82
7	47
16	53
27	95
19	91
9	60
23	87
4	46
12	63
18	71

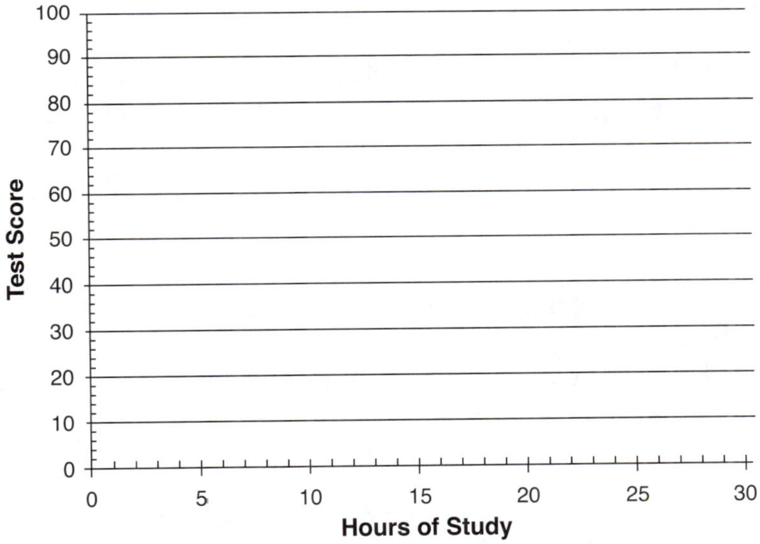

4. Does the pattern on the scatter plot show a correlation between employees' test scores and hours of study? If so, is the correlation positive or negative?

Decide whether the following groups of data would likely have a positive correlation, a negative correlation, or no correlation.

5. Math test score and mile run time

6. Temperature and ice cream sales

7. Temperature and heating bills

8. Length of investment and interest earned

Answers start on page 164.

LINE GRAPHS

A **line graph** is a graph that connects the plotted data points with a line. Line graphs are especially useful when showing how a quantity changes over time.

The line graph at the right shows the gross national income of the United States from 2005 to 2012. The gross national income is the total value of goods and services produced by a nation. The dots represent data points. From the graph, you can see that the gross national income has mostly increased over time, with only one period of decrease.

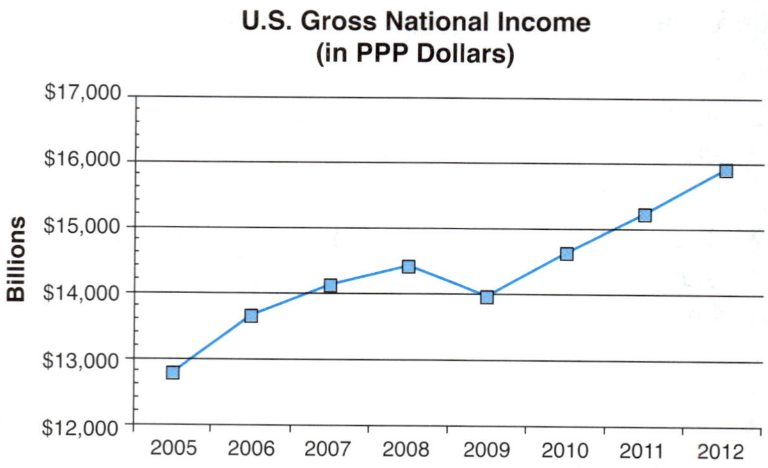

Source: World Bank

FINDING INFORMATION ON A LINE GRAPH

Example What was the *approximate* gross national income of the United States in 2008?

Step 1
Locate 2008 on the horizontal axis and find the data point directly above it.

Step 2
Read the value on the vertical axis closest to the data point.

The data point is about halfway between 14,000 and 15,000. Note that the values on the vertical axis represent billions of dollars. The gross national income in 2008 was about **$14,500 billion**.

Use the line graph above for problems 1–4.

1. In which year was the gross national income approximately $13,600 billion?

2. By about how much did gross national income increase from 2009 to 2012?

3. In which year did the gross national income increase the most?

4. In which year did the gross national income decrease?

Use the line graph below for problems 5–8.

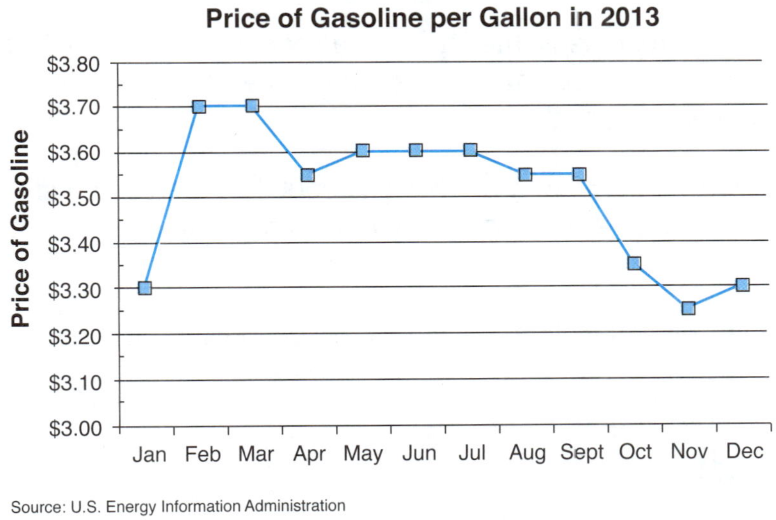

5. About how much greater was the price of gasoline in February than in January?

6. During which period did the price of gasoline remain constant longest?

7. During which periods did the price of gasoline decrease?

8. What was the overall change in price from February to November?

Use the line graph below for problems 9–12. Use the legend to find out which country each line represents.

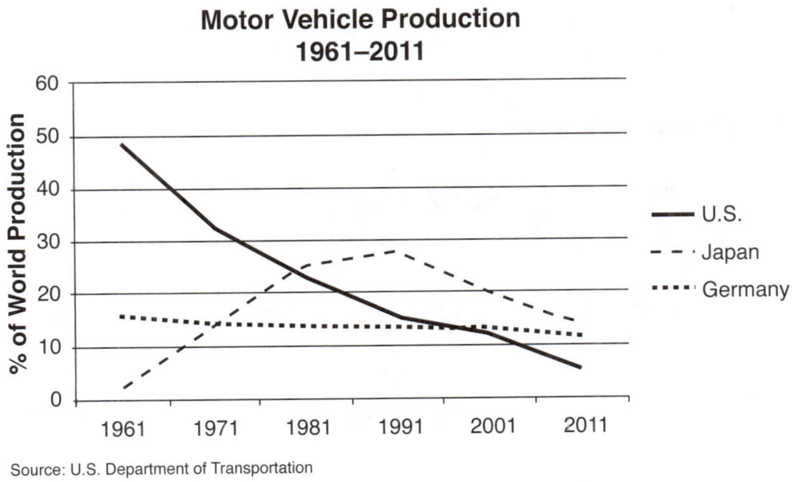

9. In which year did Japan first produce more cars than the United States?

10. In which year did Germany first produce more cars than the United States?

11. Which country's share of the market remained the most constant from 1961 to 2011?

12. Explain Describe Japan's share of the automobile market from 1961 to 2011.

Answers start on page 164.

FINDING THE MEAN

A **typical value** is an amount that represents the values in a set of numbers. A typical value must fall within the series of numbers it represents. There are three common typical values in a set of numbers: **mean**, **median**, and **mode**.

When most people use the word *average*, they are talking about the mean. For a set of values that are similar to each other, the mean is what you want to know when you ask, "About how much . . .?"

To find the mean, add the values in a set of data, and then divide the sum by the number of values in the set.

FINDING THE MEAN

Example Ella scored 14, 18, 11, and 16 points in her last 4 basketball games. What is her scoring average?

Step 1
Find the mean. Add the values in the set.

$14 + 18 + 11 + 16 = 59$

Ella's scoring average is **14.75 points**.

Step 2
Divide the sum by the number of values in the set.

$\frac{59}{4} = 14\frac{3}{4} = 14.75$

The mean is not necessarily equal to any number in the set. However, it must fall within the range of the numbers in the set.

Find the mean for each set of numbers.

1. 35, 49, 31, 25	\$1.25, \$2.08, \$0.97, \$3.12, \$1.48	5, 4, 12
2. 174, 12, 398, 214, 45	6, 3, 15, 9, 4	\$35, \$12, \$97, \$6, \$51, \$90
3. \$215, \$154, \$192	87, 96, 95, 100, 75, 82, 88	32.1, 15.75, 17.2, 18.35, 28.4
4. 5.6, 5.1, 6.4	\$240, \$198, \$264, \$249, \$220	1, 7, 13, 7, 10

Solve.

5. In July, DeShaun's weekly grocery bills were \$84, \$98, \$72, and \$60. What is his mean weekly grocery bill for the month?

6. Miranda received scores of 83, 78, 85, 82, and 77 on 5 tests. What is her average score on the 5 tests?

If you know the mean, you can use algebra to find a missing value. You can also use the mean to draw conclusions and make predictions.

USING THE MEAN

Example 1 Sean budgets \$400 a month for food. In 5 months, he spent \$415, \$380, \$489, \$392, and \$404. In order to average \$400 per month, how much should he spend in the sixth month for food?

Step 1
Write an equation to represent the mean. Let x represent the amount spent in the sixth month.

$$\frac{415 + 380 + 489 + 392 + 404 + x}{6} = 400$$

Step 2
Solve for x.

$$\frac{415 + 380 + 489 + 392 + 404 + x}{6} = 400$$
$$\frac{2{,}080 + x}{6} = 400$$
$$2{,}080 + x = 2{,}400$$
$$x = 320$$

Sean should spend **\$320** in the sixth month to average \$400 per month.

Example 2 Gina is paid weekly based on the amount of sales she makes. Her first paychecks are for the following amounts: \$346, \$358, \$320, and \$336. Based on the mean for the first 4 weeks, about how much will Gina make in a year?

Step 1
Find the mean of the first 4 weeks.

$$\frac{346 + 358 + 320 + 336}{4} = 340$$

Step 2
Gina will typically make \$340 a week. There are 52 weeks in a year. Multiply the mean by 52.

$$\$340 \times 52 = \$17{,}680$$

Gina will make about **\$17,680** in a year.

Solve as directed.

7. Jena plans to run an average of 6 miles a day. For the first 6 days, she runs 3, 8, 8, 0, 4, and 5 miles. How many miles should she run on the seventh day to meet her goal?

8. In the first 8 weeks of school, students at Pine Elementary read the following numbers of books.

   80 85 92 78 80 81 90 96

   Based on the average of the first 8 weeks, how many books will they read in a 36-week school year?

9. Jared set a goal to sell 300 alarm systems in a 60-day period. At the end of the 60 days, he had averaged 4 alarms per day. Did he meet his goal?

10. **Explain** Carmel received test scores of 77, 81, 95, and 91. She has 1 more test to take. In order to get an A in the class, she needs an average of 90 on the 5 tests. If the highest score Carmel can earn on the last test is 100, is it possible for her to get an A in the class? Explain your thinking.

Answers start on page 164.

MEDIAN AND MODE

There are other typical values that might be more representative of a set of data. The **median** is the middle value of a set of data. To find the median, sort the data from least to greatest. For an odd number of values, the middle number is the median. For an even number of values, the median is the mean of the two middle numbers.

FINDING THE MEDIAN

Example Angela has helped 5 dogs onto the examination table. Their weights are 40 lb, 10 lb, 140 lb, 35 lb, and 30 lb. What is the median weight of the dogs?

Step 1
Sort the numbers numerically from least to greatest.

10 lb 30 lb 35 lb 40 lb 140 lb.

The median weight is **35 lb**.

Step 2
Choose the middle number from the set. (If there is an even number of values, find the mean of the two middle values.)

Another typical value is the **mode**—the value in the set that occurs most often. (If each value occurs only once, there is no mode in the set.)

FINDING THE MODE

Example A restaurant offers these prices on complete meals: Fried Chicken—$8.99, Steak—$10.99, Fish—$8.99, Shrimp—$10.99, Lasagna—$8.99, Hamburger—$7.99. What is the mode for these prices?

Count how many times each value occurs.

$7.99: 1 time $8.99: 3 times $10.99: 2 times

Because it occurs the most (3 times), **$8.99** is the mode.

Find the median.

1. $18, $75, $4, $23, $51 53, 22, 12, 18, 50 13.4, 8.6, 4.1, 80.9

2. 0.8, 1.4, 2.3, 0.6, 1.6 62, 78, 73, 45, 67, 90 $457, $375, $223, $480

Find the mode.

3. 35, 18, 21, 21, 18, 97, 32, 18 22°, 17°, 31°, 22°, 18°, 35°, 22°, 31°

4. $6.50, $5.75, $6.00, $5.50, $6.75 23, 24, 26, 24, 28, 23, 25, 23, 27, 25, 24

Sometimes, one or more data values are very large or very small compared to the other values in the set. These extreme values are called **outliers**. Outliers don't affect the median as much as they affect the mean.

USING THE MEAN AND MEDIAN WITH OUTLIERS

Example A jewelry store has 5 different rings on sale for \$27, \$26, \$33, \$31, and \$108.

Your common sense tells you that \$30 is a typical price for a ring. Compare how the outlier of \$108 impacts the mean more than the median.

Find the mean of the 5 rings:

\$27 + \$26 + \$33 + \$31 + \$108 = \$225
\$225 ÷ 5 = \$45

The mean, \$45, is much higher than \$30.

Find the median of the 5 rings:

\$26 \$27 \$31 \$33 \$108
The middle value is \$31.

The median, \$31, is very close to \$30.

Notice in the example above that while the mean may not be a typical value, the median is.

First, identify the outlier. Then find the mean and the median of the data set.

5. 13.4 kg, 14.2 kg, 3.2 kg, and 12.8 kg

Outlier: ___________

Mean: ___________

Median: ___________

6. \$126,000; \$74,000; \$86,000; \$85,000; and \$84,000

Outlier: ___________

Mean: ___________

Median: ___________

7. 61 in, 68 in, 44 in, 60 in, 69 in, and 64 in

Outlier: ___________

Mean: ___________

Median: ___________

8. 73, 69, 76, 107, 80

Outlier: ___________

Mean: ___________

Median: ___________

Solve.

9. The following data shows the number of sales made in a month by 8 out of 9 salespeople.

45 34 31 48 37 43 32 42

The 9th employee made 66 sales in that month. By how much does the mean change when the 9th value is added?

10. **Explain** The data below gives the margin of victory for a high school football team's games in a season.

3 7 21 14 3 17 30 3

Do you think the mean, median, or mode is the most reliable typical value for the data? Explain your reasoning.

Answers start on page 164.

Problem Solver

USING TYPICAL VALUES

When news sources report the results of a survey or study, they don't usually tell you all of the data. Instead, they might sum up the study with a typical value. You must use your knowledge of mean, median, and mode to interpret the information.

Here are some useful tips to remember:

- When people say "average," they're usually talking about the mean. Be careful: outliers can make the mean misleading.
- The median is the middle value. Outliers don't have a big effect on the median.
- The mode is the value that occurs most often. The mode is useful when you care what is popular or what happens frequently.

Use your knowledge of mean, median, and mode to answer the following questions.

1. In Mr. Bound's class, 1 student scored a 76 on a test, 10 students scored above 76, and 10 students scored below 76. What typical value does a score of 76 represent?

A. mean

B. median

C. mode

2. The manager at a toy store wants to reorder the items that sell the best. Which typical value should he use to decide which items to reorder?

A. mean

B. median

C. mode

3. **Explain** You are researching housing prices. You can afford a house that will sell for about $150,000. You know the mean, median, and mode for a certain neighborhood. Which typical value would be most helpful in deciding whether you can afford a house in this neighborhood? Explain your answer.

4. An office poll asks employees how many cups of coffee they drink in a week. The mean of the data is 15, but the median is 21. Which of the following is true?

A. A few people drink much fewer than 21 cups of coffee a week.

B. A few people drink much more than 21 cups of coffee a week.

C. Most people drink between 15 and 21 cups of coffee a week.

5. Out of the 25 students in a 5th grade class, 22 are between 49 and 51 inches tall. Three students are 59 inches tall. Which of the following must be true?

A. The median is higher than the mean.

B. The mean is higher than the median.

6. **Explain** The mean salary for NBA basketball players is about $5 million a year. The median salary for NBA players is about $2.5 million a year. Why do you think the two typical values are so different?

Sometimes line graphs display the mean, median, or mode of a data set. These graphs can show how typical values change over time.

USING TYPICAL VALUES TO INTERPRET GRAPHS

Example This graph shows the U.S. median age from 1970 to the estimated value for 2020. What can you conclude about the U.S. population?

You can see from the graph that the U.S. median age has increased from 1970 to the present. But what does the median say about the whole population?

The median is the middle value. In 2010, the median age was 37. This means that the number of people younger than 37 was equal to the number of people older than 37. You can imagine there are many people younger than 37, so there must be many people older than 37 as well.

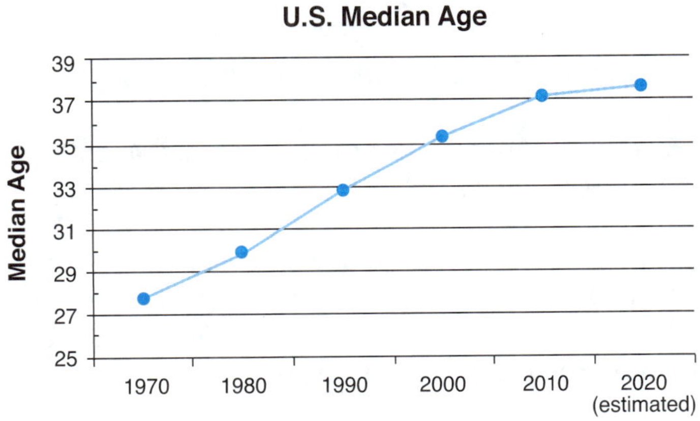

Source: U.S. Bureau of the Census

If the median age in the United States is rising, then you can conclude that the population as a whole is getting older.

Use the bar graph for problems 7 and 8.

The graph shows the average high temperature in a forest from March to June. Researchers recorded the high temperature for each day in a month and calculated the average of those values.

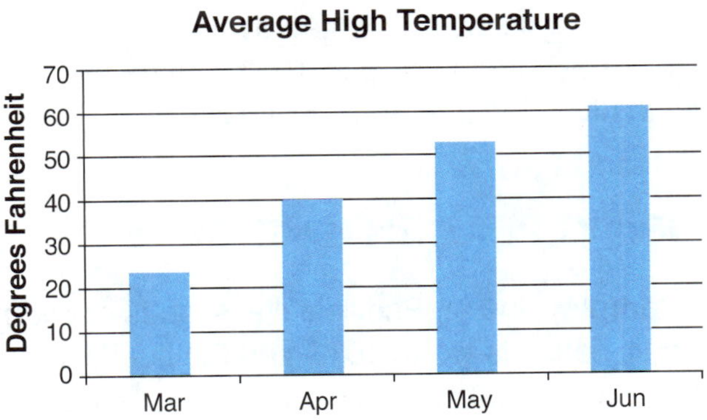

7. Which typical value does this graph use?

8. On April 28, the high temperature was 51°F. What can you conclude about other days in April?

Use the circle graph for problems 9 and 10.

The graph shows the breakdown of dress sales at a boutique over one weekend.

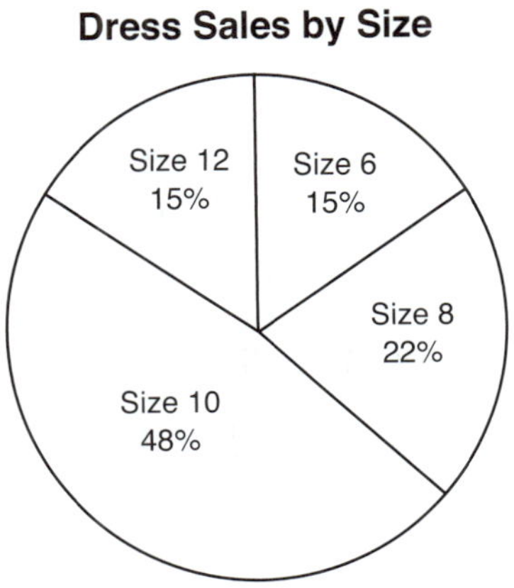

9. Which size represents the mode of the data?

10. **Explain** Can you determine from the graph what the median size would be? Explain your answer.

Answers start on page 164.

WEIGHTED AVERAGES

Sometimes, the values in a data set may not be equally important. For example, a final exam might count as two tests. If you were finding your average in the class, your final exam score would be worth more than your other scores.

When some values in a data set are worth more than others, you must find the **weighted average**.

FINDING A WEIGHTED AVERAGE

Example Aparna's average for the first 4 tests is 80. Her score on the fifth test is 90. What is her average in the class?

You might be tempted to say 85 because it is the mean of 80 and 90. Look at the diagram to the right. Note that 80 represents 4 tests, but 90 represents only 1. You must find the weighted average.

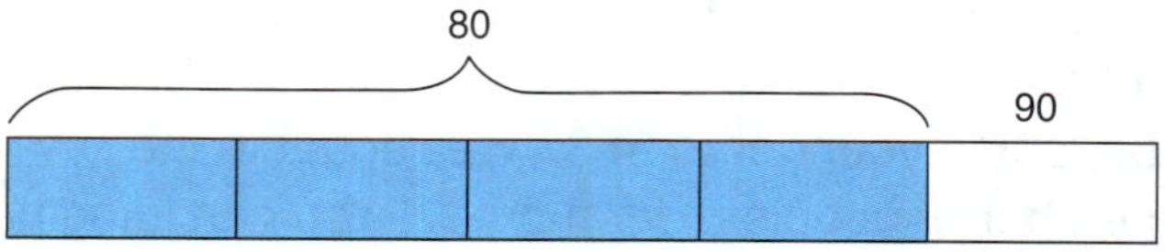

Step 1
Multiply each score by the weight. Then add.

$(80 \times 4) + (90 \times 1) = 410$

Step 2
Divide by the number of values. Aparna took 5 tests.

$410 \div 5 = 82$

Aparna's average is **82**.

Some averages are calculated by assigning percentages to different categories. These percentages must add up to 100%. Since the percentages represent one whole, you don't need to divide by the number of values.

USING PERCENT WEIGHTS

Example In Ms. Padilla's class, tests are worth 30% of the final grade, homework is worth 20%, participation is worth 10%, and the final exam is worth 40%. Jeff earns an 80 on the tests, 95 on homework, 100 on participation, and an 85 on the final exam. What is his average in the class?

Step 1
Multiply Jeff's scores by the percent weights.

Tests:	$80 \times 0.3 = 24$
Homework:	$95 \times 0.2 = 19$
Participation:	$100 \times 0.1 = 10$
Final Exam:	$85 \times 0.4 = 34$

Step 2
Add.

$24 + 19 + 10 + 34 = 87$

Jeff has an average of **87** in the class.

Solve as directed.

1. On a recent road trip, Cesar averaged 65 mph for the first 4 hours and 80 mph for the final 6 hours. What was his average speed for the entire trip?

2. An online food critic rates restaurants based on four criteria. She assigns the following weights to the categories.

Category	Weight
Taste	35%
Presentation	20%
Service	25%
Atmosphere	20%

Blossom Café receives the following:
8 for Taste, 7.8 for Presentation, 8.4 for Service, and 8.2 for Atmosphere. What is Blossom Café's total weighted score?

3. The table shows how many hours Yeri worked each day during the past three months.

Number of Hours	Number of Days
3	3
6	18
8	54
11	15

What is the average number of hours Yeri worked per day over the past three months? Round your answer to the nearest tenth. (***Hint:*** Treat the number of days as the weight.)

4. In the first 10 games of the season, a quarterback has a 60% pass completion rate. In the last 5 games, his pass completion rate is 78%. What is his pass completion rate for the entire season?

5. At a gymnastics meet, Michelle receives 152 points for execution and 110 points for artistry. Execution makes up 65% of her total score, while artistry makes up 35% of her score. What is Michelle's final score?

6. Mr. Reyes gives his students the following circle graph to explain their final grade.

Final Grade Breakdown

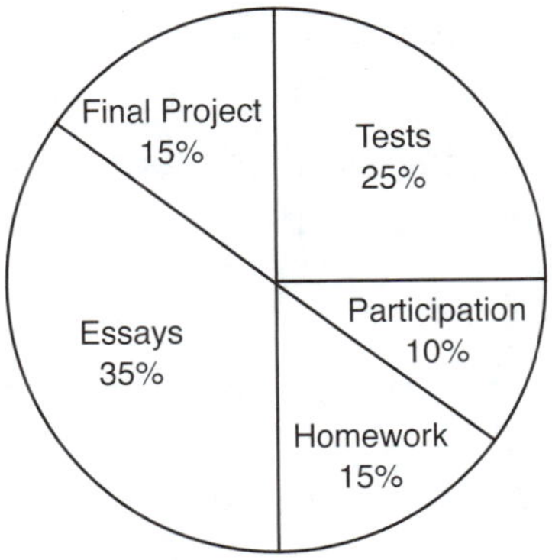

Christian receives a 75 for tests, a 78 for participation, an 85 for homework, an 80 for essays, and a 90 for the final project. What is his final grade in the class, rounded to the nearest whole number?

7. Paul buys two suits for $175 each at one store and three suits for $150 each at a different store. How much did Paul pay on average per suit?

8. After the first three months of the season, a baseball player's batting average is .300. After another month, his average for the four months is .275. Which of the following is most likely?

A. He did not play the last month.
B. He batted about .200 the last month.
C. He batted about .300 the last month.
D. He batted about .350 the last month.

Answers start on page 165.

Mixed Review

Problems 1–4 are based on the bar graph.

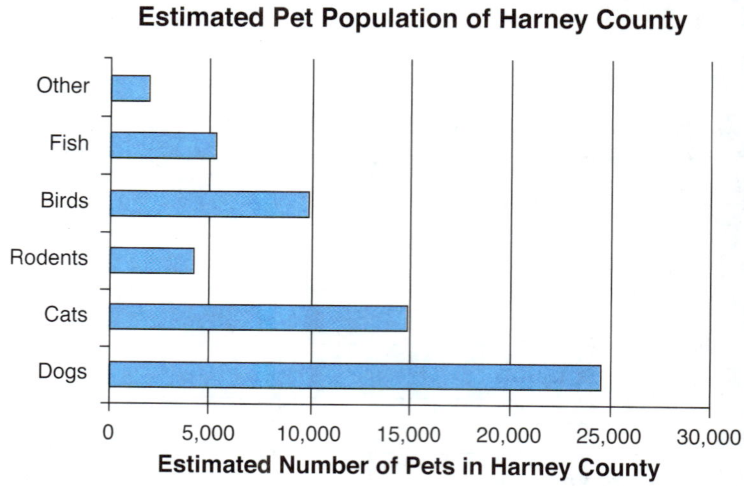

1. About how many cats are in Harney County?

2. There are about half as many fish as which other group of pets?

3. About how many more dogs are there in Harney County than cats?

4. There are about two-thirds as many birds as which other group of pets?

Problems 5–8 are based on the scatter plot.

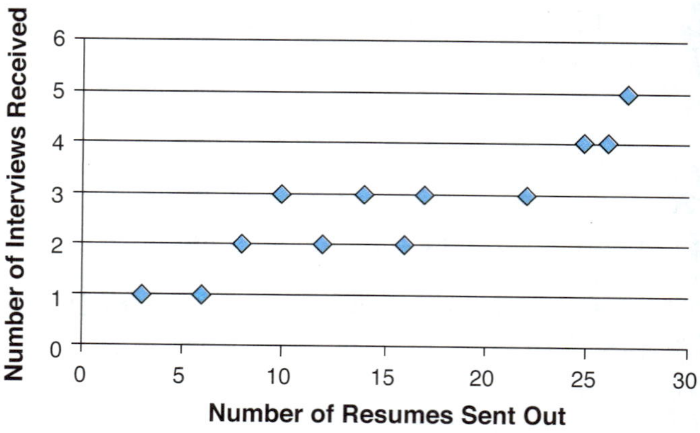

5. How many people participated in the survey graphed on the scatter plot?

6. What fraction of the participants received 3 or more job interviews?

7. Describe the correlation, if any, between the number of resumes sent out and the number of interviews received.

8. Based on the scatter plot, if a person sent out 20 resumes, about how many interviews do you think the person will receive?

Problems 9–12 are based on the line graph.

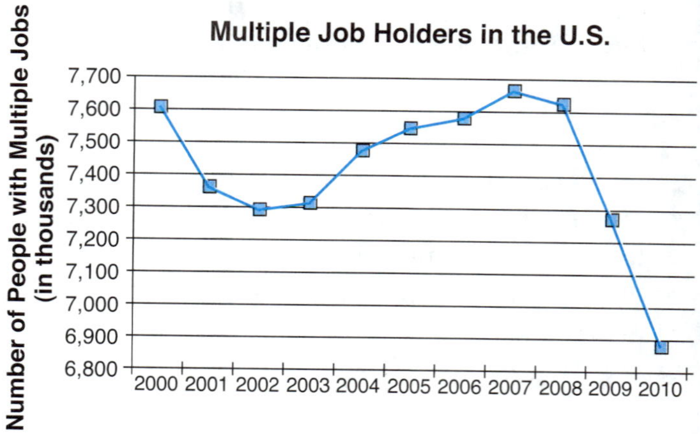

9. About how many people held multiple jobs in 2000?

10. In 2009, the number of multiple jobholders dropped to about the same level as which other year?

11. Estimate the mean number of multiple jobholders from 2007 to 2010. Round to the nearest whole.

12. From 2007 to 2010, what was the median number of multiple jobholders in the United States?

Problems 13–16 are based on the histogram.

Length of Eruption of Old Faithful, July 21

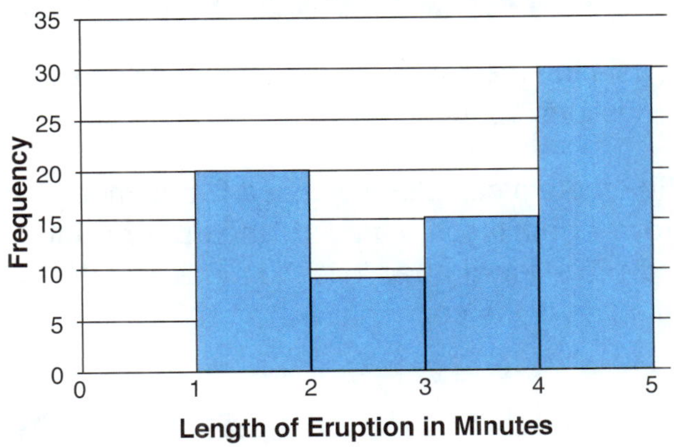

13. How many geyser eruptions lasted between 1 and 2 minutes?

14. How many eruptions of Old Faithful does this graph represent?

15. Did more geyser eruptions last longer than 3 minutes or shorter than 3 minutes?

16. What percent of geyser eruptions lasted between 3 and 4 minutes? Round your answer to the nearest percent.

Problems 17–19 are based on the circle graph.

Sources of Funds for Research and Development (U.S., 2008)

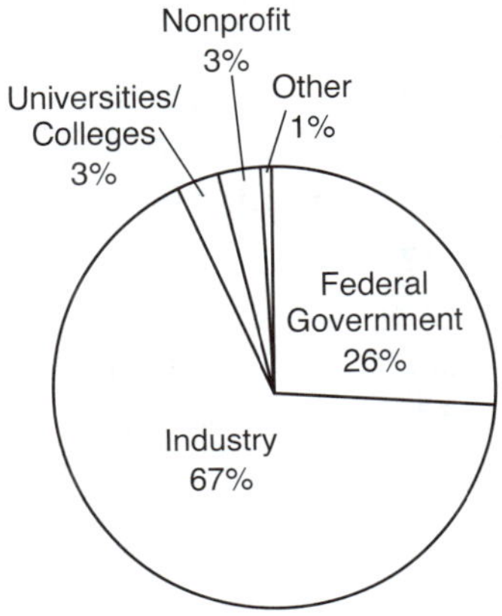

Source: Statistical Abstract of the United States, 2012

17. In 2008, what was the largest source of funds for research and development in the United States?

18. Out of each $1 million given to Research and Development, how much came from the federal government?

19. About what fraction of funding comes from sources other than industry?

Solve as directed.

20. The following line plot shows the weights of packages that are brought to the post office in a day. Find the mean, median, and mode of the data.

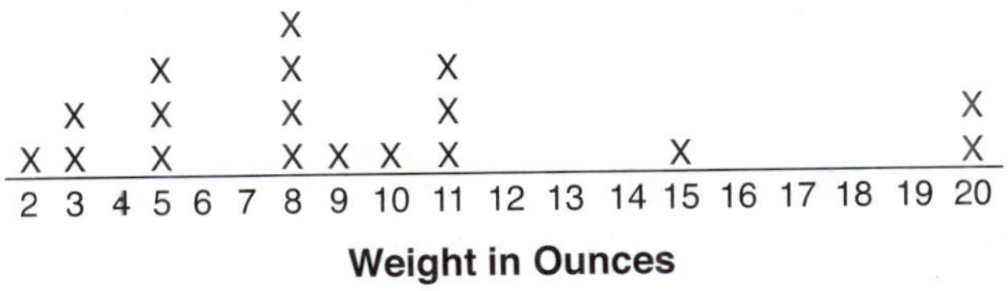

21. The prices of laptops at an electronics store are shown in the table.

Price	Number of Laptops
$800	2
$1,500	5
$2,750	4
$3,900	1

What is the mean price of laptops at this store?

22. Explain On a recent math test, most of the 27 test scores were above a 70. However, 4 students received a 30 on the test. How reliable would the mean of the test scores be? Which typical value would you use to represent the data?

Answers start on page 165.

READING A BOX PLOT

While the median tells you about the center of a data set, you can use a **box plot** to answer questions about an entire set of data.

To create a box plot, you need five numbers: the **minimum** (the lowest value in the set), the **lower quartile**, the median, the **upper quartile**, and the **maximum** (the highest value in the set). The lower quartile is the median of the lower half of the data. The upper quartile is the median of the upper half.

FINDING THE UPPER AND LOWER QUARTILES

Example Ms. Rivera recorded the following test scores. Find the upper and lower quartiles of the data.

55 63 68 76 79 80 80 81 83 93 94 98

Step 1
Find the median. Make sure the data is arranged in order from least to greatest.

Of the 12 values, the two middle values are 80. The median is 80.

Step 2
Split the data in half at the median. Find the median of the lower half.

55 63 68 76 79 80

68 + 76 = 144
144 ÷ 2 = 72

The median of the lower half is 72.

Step 3
Find the median of the upper half.

80 81 83 93 94 98

83 + 93 = 176
176 ÷ 2 = 88

The median of the upper half is 88.

The lower quartile is **72**. The upper quartile is **88**.

A box plot is drawn on a scale. Draw a box that starts at the lower quartile and ends at the upper quartile. Draw a vertical line inside the box at the median of the entire data set. Then draw lines extending to the minimum and the maximum. These lines are called **whiskers**. This kind of chart is also called a **box-and-whisker plot**.

Look at the box plot of the test scores.

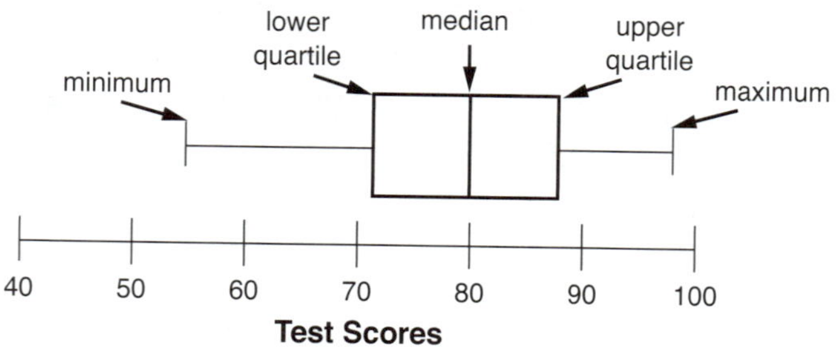

Each section of the plot represents 25% of the test scores.

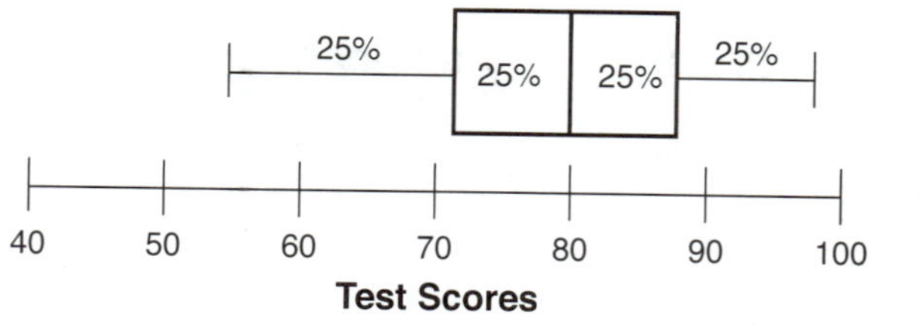

Students in the *bottom 25%* of Ms. Rivera's class scored between 55 and 72. Students in the *top half* of her class scored between 80 and 98.

Find the upper and lower quartiles for the following data sets.

1. 1, 1, 2, 3, 3, 5, 6, 7, 8, 9, 10, 11, 15

2. 4.2, 4.8, 5.1, 5.3, 5.3, 5.7, 6.1, 6.2

3. 23, 41, 52, 27, 33, 38, 20, 25, 39, 47, 46

4. 4, 76, 28, 52, 101, 31, 44, 63, 92, 89

Use the box plot for problems 5–8.

5. Which age represents the lower quartile?

6. What is the median of the data?

7. What percent of gym members are over 40 years old?

8. If the gym has 400 members, how many are between the ages of 25 and 40?

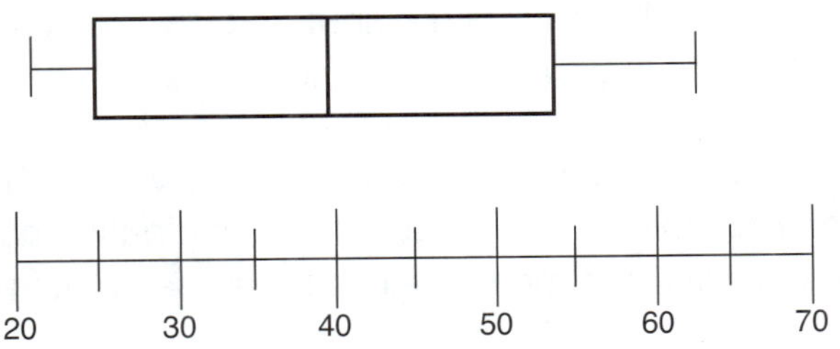

Age of Gym Members

Use the double box plot for problems 9–12.

9. Which highway has a higher median speed?

10. What is the difference between the two highways' maximum speeds?

11. **Explain** What percent of cars drive between 60 and 70 mph on each highway? Can you conclude that more cars drive between 60 and 70 mph on the highway with the greater percentage?

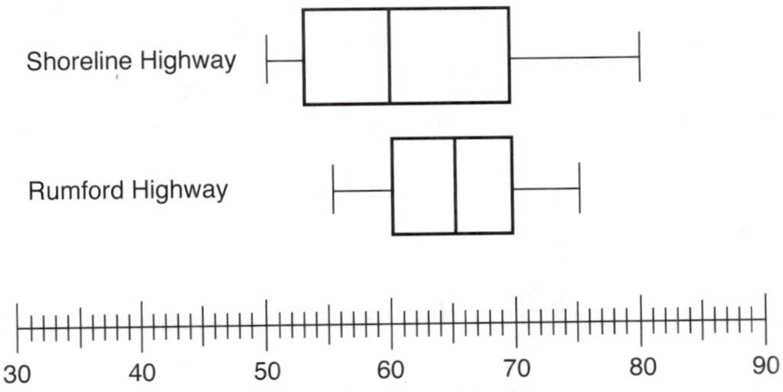

Highway Speeds (mph)

Construct a box plot using the data below.

12. Monthly sales numbers:

17	13	8	20	24	30
23	27	11	26	10	18

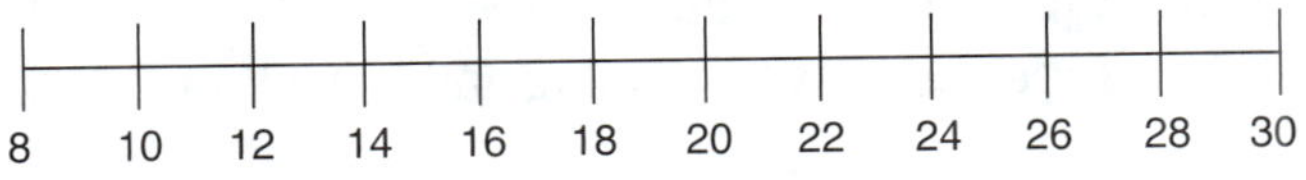

Monthly Sales

Answers start on page 166.

DISTRIBUTION OF DATA

When you represent data in a graph or plot, you can examine the **distribution**, or shape, of the data. It's easier to think about distribution when you draw a curve that represents the data.

Look at the histogram to the right. A curve has been drawn on top of the data. This specific curve is called the **normal distribution curve**.

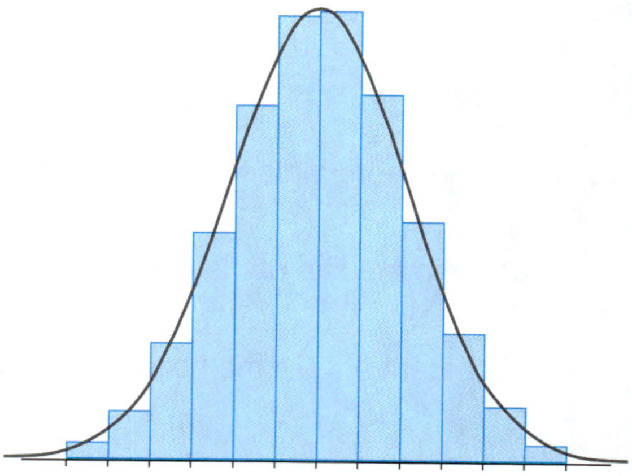

Data that fits the normal distribution curve is **symmetric**. The left side looks the same as the right side.

You can see from the histogram that the values that occur most often are in the center. On a normal distribution curve, the mode, the mean, and the median all occur at the peak in the center.

When a data set has outliers, the curve becomes **skewed**.

When data has low outliers, it is skewed to the left. The mode is still at the peak, but the mean and the median have been pulled to the left.

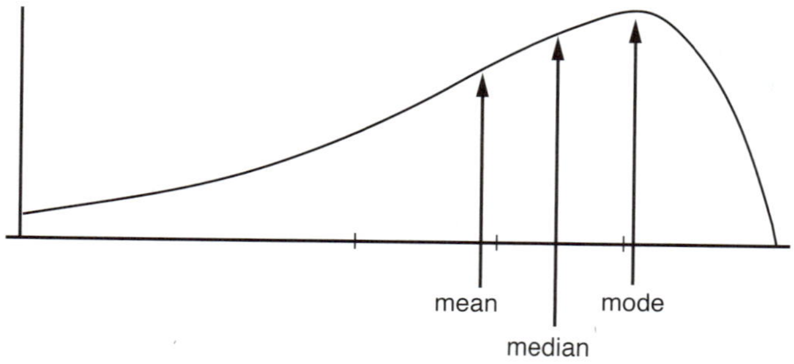

When data has high outliers, it is skewed to the right. The mean and the median have been pulled to the right.

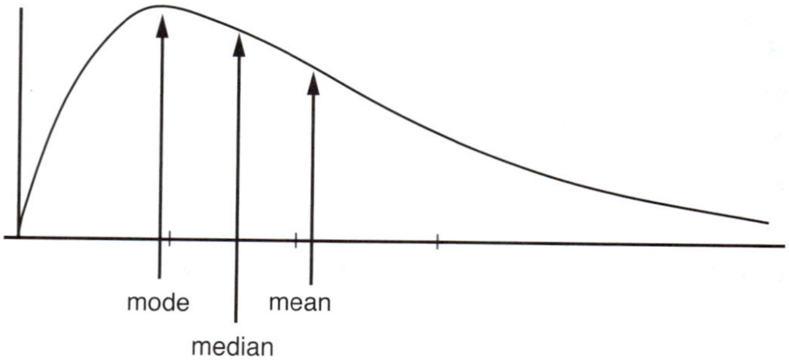

You can also describe distribution by observing **variation**.

Data with **low variation** is clustered together.

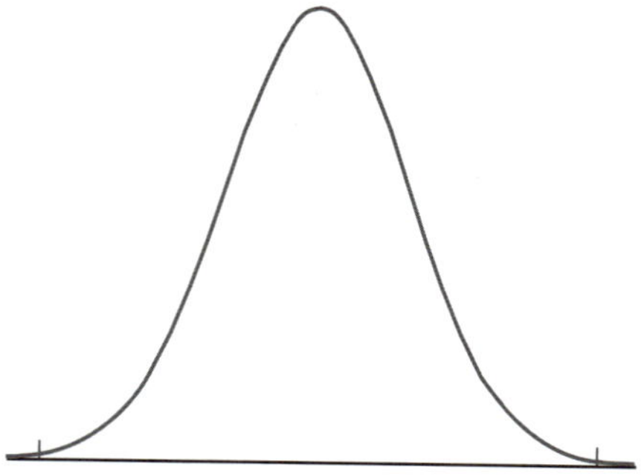

Data with **high variation** is spread out.

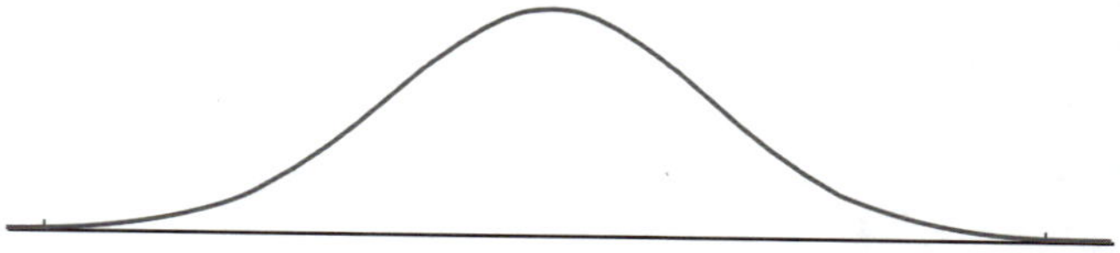

When you examine variation, it is often useful to know the **range**. The range is the difference between the highest value and the lowest value in a data set. Data with low variation will have a small range. Data with high variation will have a large range.

Histograms and box plots are helpful when studying distribution. On a histogram, you can see the shape of the curve. On a box plot, the data is skewed if the box looks off-center. Box plots with high or low variation might seem stretched or compressed.

Use the histogram for problems 1–3.

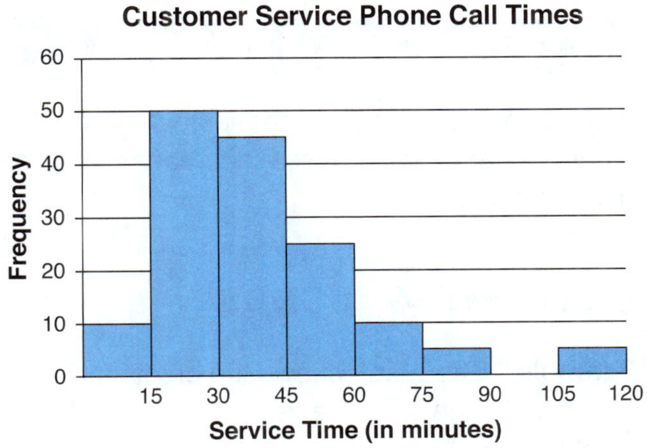

1. Is the data skewed or symmetric? If the data is skewed, is it skewed left or right?

2. Which time interval occurs most frequently? What conclusion can you draw about customer service times?

3. Does the data have any outliers? Which values could be considered outliers?

Use the box plot for problems 4–8.

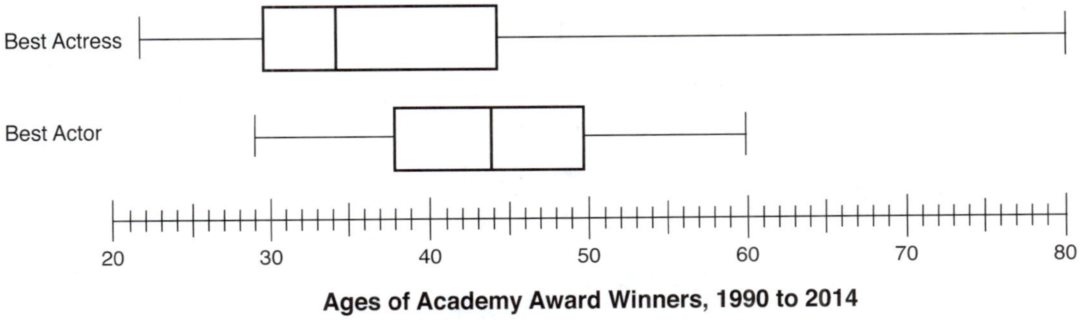

Source: Academy of Motion Picture Arts and Sciences

4. What is the range of ages for Best Actress winners from 1990 to 2014? For Best Actor winners?

5. Which data set has greater variation, Best Actress or Best Actor?

6. Do you think either data set has outliers? Where do you expect the outliers are located?

7. The data for Best Actress winners is skewed. Which direction is it skewed?

8. **Explain** Describe the distribution of the data for Best Actress and Best Actor. What differences are there in the typical ages of the actors and actresses who receive the award?

Solve as directed.

9. The minimum and maximum values in a data set are 31 and 95. The mode is 67, the median is 62 and the mean is 52. Is this data symmetric, skewed right, or skewed left?

10. Which data set would have higher variation: driving times through downtown during rush hour or driving times in the countryside at 4 A.M.?

11. In Major League Baseball, the salaries of the top players are much higher than most other players. Is the distribution of baseball salaries symmetric, skewed right, or skewed left?

12. The distribution of grades on a test is skewed right. If you want a high grade, do you want your grade to be closer to the mean, median, or mode?

Answers start on page 166.

MAKING PREDICTIONS AND IDENTIFYING TRENDS

Trends show how situations change over time. When you look at a graph or a table, observe whether the data set increases, decreases, or remains steady. If you see a pattern, you have identified a trend.

IDENTIFYING TRENDS

Example The graph shows the profits for Intricate Design over five years. What trends does the graph show?

Look for places where the graph increases, decreases, or remains steady. This graph shows two trends.

Trend #1: Profits decreased sharply from Year 1 to Year 3.

Trend #2: Profits increased slightly from Year 3 to Year 5.

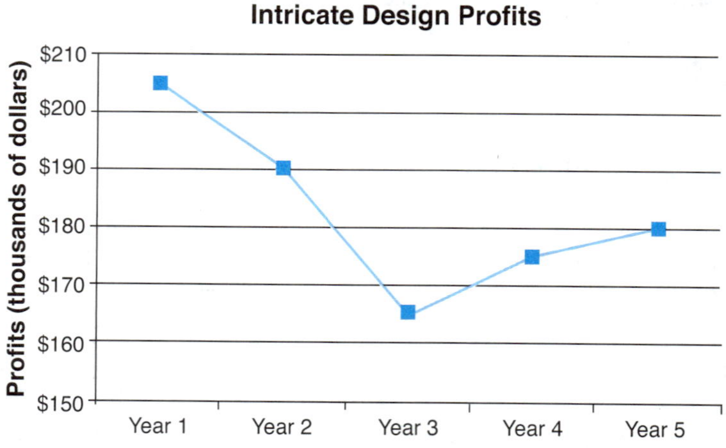

You can use graphs and tables as the basis for making **predictions**. Predictions use information from past events to guess what will happen in the future.

To make a prediction, you need to determine the trends in a data set and then decide if you think the current trend will continue.

MAKING PREDICTIONS

Example If Intricate Design profits are more than $190,000 in Year 6, it will open a new storefront. Based on the graph above, are profits likely to be over $190,000?

To make a prediction about the future, look at current trends. From Year 3 to Year 5, profits increased slightly. To exceed $190,000 in Year 6, profits would have to increase by more than $10,000. Look closely at the graph to see if that's likely to happen.

In Year 4, profits increased by about $10,000. In Year 5, profits increased by only about $5,000. This suggests that Intricate Design profits are starting to level off. Profits will probably not be greater than $190,000 in Year 6.

You can make predictions when a graph shows change. Line graphs, tables, and bar graphs can all be used to show trends.

Use the line graph for problems 1–3.

1. Describe the trends in the graph of customers leaving MPE Networks.

2. Describe the most current trend.

3. **Explain** Based on the data, how many customers do you predict left MPE Networks in 2014?

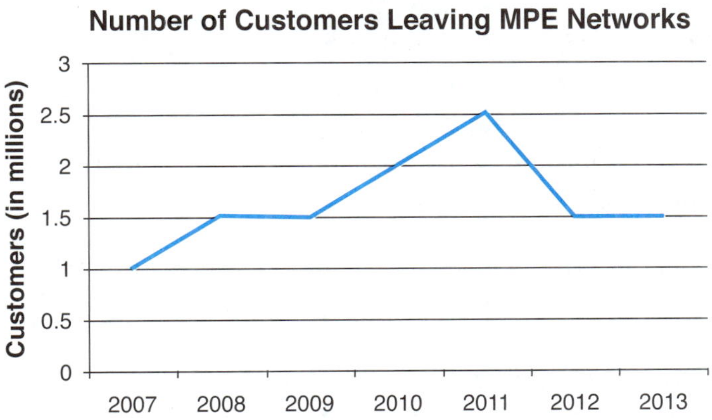

Use the bar graph for problems 4–6.

4. What trends do you see in the sale of tickets to R-rated movies at the Dollar Theater?

5. Do you think the current trend will continue? Why or why not?

6. Based on the current trend, what types of movies would you recommend the Dollar Theater show in the future?

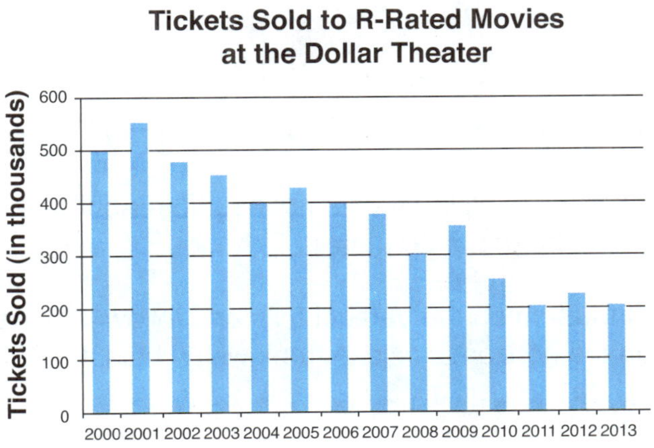

Use the table for problems 7–10.

The table shows the average stock price per month for three different stocks.

	Jan	Feb	Mar	Apr	May
RAO	2.10	2.14	2.15	2.19	2.23
LEC	5.43	5.44	5.46	5.45	5.46
KBW	11.07	11.25	11.01	10.88	10.54

7. What trends do you see in
   a. the price of RAO?
   b. the price of LEC?
   c. the price of KBW?

8. What kind of graph would you use to best show the trends in the table?

9. Based on the current trends, estimate the price of each stock in June.

10. **Explain** Could you use the table to make predictions about stock prices for the rest of the year? Explain your answer.

Answers start on page 166.

UNDERSTANDING CORRELATION

You've already learned that data can have a positive or negative correlation. Groups of data are said to have a **linear correlation** when the points on the scatter plot approximately form a line.

The scatter plot to the right compares the diameter and height of tree trunks.

This data has a linear correlation because all of the plotted points are close to the black line.

The black line is called the **line of best fit**. This line estimates the relationship between the two groups of data.

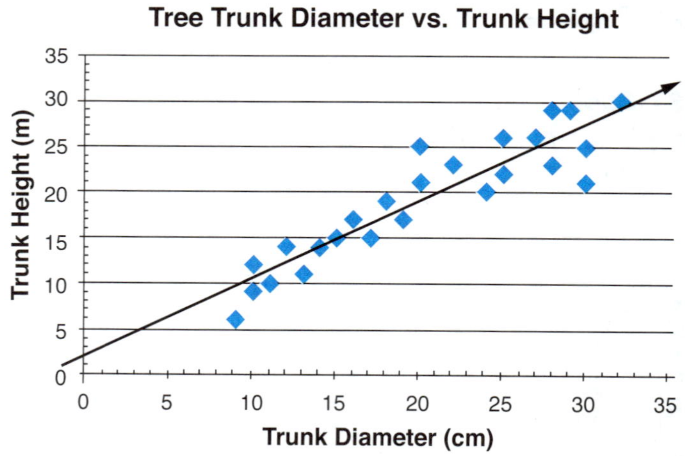

USING THE LINE OF BEST FIT

Example Using the line of best fit graphed above, find the trunk diameter for a tree with a height of 20 meters.

Find 20 m on the vertical axis and follow it to the black line. Read the corresponding value on the horizontal axis. A trunk with a height of 20 m probably has a diameter of about **21 cm**.

Some groups of data have a **nonlinear correlation**. Even though the points don't form a line, you can still find a relationship between the groups of data.

ANALYZING NONLINEAR CORRELATION

Example A beverage company does a taste test to determine how much added sugar produces the most popular taste. How much sugar results in the best taste?

From 0 to 8 teaspoons, sugar and taste have a positive correlation. More sugar seems to be good.

From 8 to 11 teaspoons, sugar and taste have a negative correlation. Too much sugar ruins the drink.

The highest results in the taste test are for **8 teaspoons** of sugar per gallon. The company shouldn't add more than 8 teaspoons of sugar.

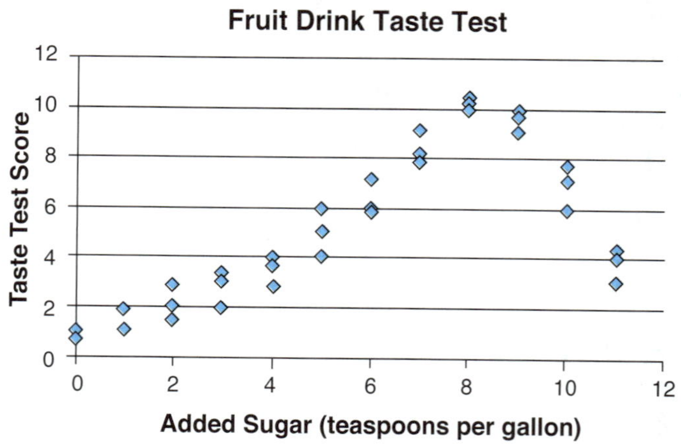

Use the scatter plot for problems 1–3.

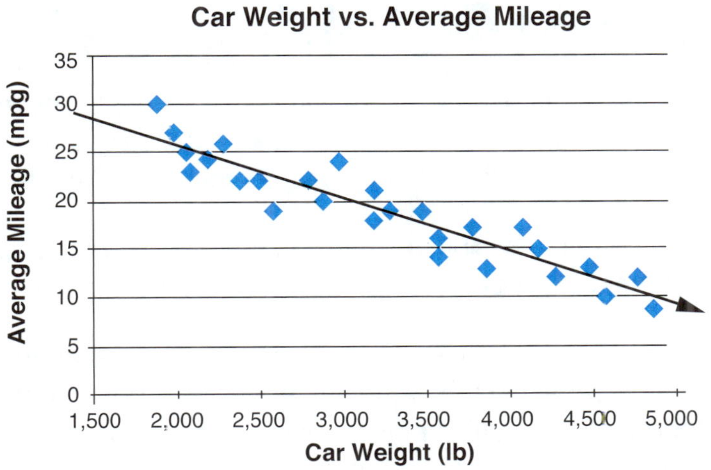

1. Describe the correlation between car weight and average mileage.

2. Based on the line of best fit, what would you expect the average mileage to be for a car weighing 3,000 pounds?

3. Tim's car weighs 2,500 pounds and averages 20 mpg. Does his car fall above or below the line of best fit?

Use the scatter plot for problems 4–6.

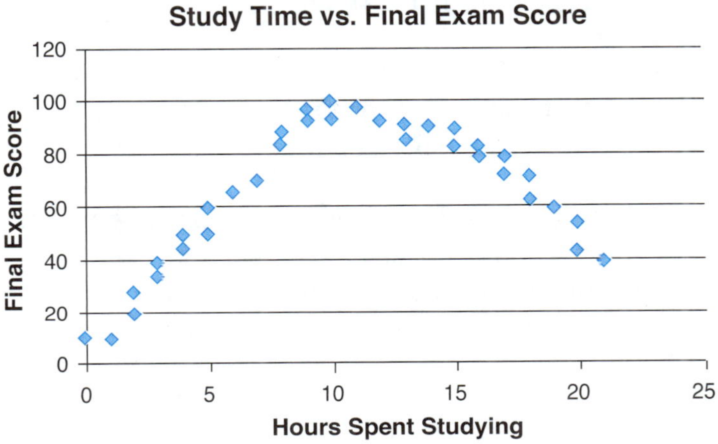

4. When is the correlation between study time and final exam score positive? When is it negative?

5. Based on the data, how long should you study to get the highest grade?

6. **Explain** Why do you think this data is nonlinear? What are some reasons that greater study time doesn't always mean a higher grade?

CORE CONNECTIONS: Correlation and Causation

A study done at an elementary school shows that students with a larger shoe size know more math. These two groups of data have a correlation. Does this mean that having big feet *causes* students to be better at math? That doesn't make sense. In elementary school, students with larger feet are older, so they have been taught more math. There is a correlation between shoe size and math knowledge, but not causation.

When you analyze correlation, use your common sense to determine whether one data group might cause the outcome you observe.

Decide whether you think the following correlations also have causation.

1. When ice cream sales increase, homicide rates also increase.

2. When the speed limit decreases, the number of accidents also decreases.

3. When people consume more calories, they gain more weight.

4. When umbrella sales are high, the number of traffic accidents is also high.

Answers start on page 167.

SAMPLING A POPULATION

Researchers often want to study a large group. The large group is called a **population**. For example, a politician might take a poll to see how many people will vote for him. The population would be the registered voters for that area.

It is often impossible to collect data from an entire population. Data collectors usually get information from a **sample population**, a small part of the larger group.

Researchers use data from the sample population to draw conclusions about the population. This means that the sample needs to accurately represent the overall population. If a politician polls only homeowners, the data won't accurately represent the entire population. Sampling works only when you have a **representative sample**.

RECOGNIZING A REPRESENTATIVE SAMPLE

Example A city councilman wants to find out what people in his district think about giving a business permit to an auto repair shop. To find out, his staff surveys 200 residents in the blocks surrounding the proposed location. Is the sample representative?

Check for **bias**. Bias skews the results and makes a sample inaccurate. In this example, the population is the entire district, but the sample includes only residents who live near the location. People who don't live near the repair shop may have different opinions. This sample is biased, so it isn't representative.

DRAWING CONCLUSIONS USING SAMPLES

Example A supermarket manager wants to know how customers feel about a proposed redesign of the store. To find out, she asks shoppers about the proposed change. She talks with shoppers at different times on different days and in different parts of the store. Her results are as follows:

Number polled: 200
Number in favor: 150
Number against: 50

Is this sample representative? What percentage of the customers can the manager reasonably expect to like the redesign?

The manager talks to random shoppers at different times, on different days, and in different areas of the store. This sample seems **unbiased and representative**.

Use the sample to find the percent in favor of the redesign: $\frac{150}{200} = 75\%$. The manager can reasonably expect about **75%** of all customers to like the change.

For each group below, identify the sample population and the population. Then tell whether the sample is representative of the population. Explain your answer.

1. Between 2:30 p.m. and 3:30 p.m., a supervisor checks the rate at which work is being done along a production line. He wants to see if the crew is working efficiently throughout the day.

   Sample: ____________________

   Population: ____________________

   Is it representative?

2. A pickup manufacturer surveys 2,000 owners of the company's new pickup. Owners are chosen at random from every state. The manufacturer wants to determine the level of owner satisfaction.

   Sample: ____________________

   Population: ____________________

   Is it representative?

3. A candidate for mayor polls the residents of every 5th apartment in an apartment complex. She wants to find out how these residents feel about the way this city-owned complex is managed.

   Sample: ____________________

   Population: ____________________

   Is it representative?

4. A customer returns a package of stale cookies to Ben's Market. The customer concludes that all packaged cookies at Ben's Market are probably stale.

   Sample: ____________________

   Population: ____________________

   Is it representative?

Solve as directed.

5. Ella surveys 150 band members to determine about how many hours a day students at her high school spend texting. Which of the following is true about Ella's survey?
   A. The sample population is random.
   B. The sample is unbiased.
   C. The sample size is too small.
   D. The sample is not representative.

6. You want to compare the food prices at Value Supermarket with the prices at other stores in town. How would you choose a representative sample?

7. Officials at a state college want to know the percentage of freshmen who are vegetarians. Which of the following samples is most likely to be representative?
   A. 100 freshmen drama majors
   B. 100 freshmen athletes
   C. 100 freshmen in a single dormitory
   D. 100 freshmen veterinary students

8. Art wants to determine what percent of people who swim at the Aquatic Center think the water temperature is too cold. How could he choose a representative sample?

Answers start on page 167.

USING TWO DATA SOURCES

Sometimes information from more than one source of data is needed to answer a question. A table, graph, or text may contain only part of the information you need.

Suppose Greg plans to drive from Los Angeles to San Francisco, but he needs to rent a car. Greg knows that gasoline costs about $3.70 per gallon. What would be the cost of driving the various rental cars based on distance, gas mileage, and gas prices?

Both the chart and graph below contain information that can be used to answer the question. The mileage chart shows the distance between the two cities. The bar graph shows the number of miles per gallon each type of rental car gets.

	Atl	Bos	Chi	Dal	LA	NO	SF
Atl		1,104	720	781	2,217	467	2,595
Bos	1,104		985	1,770	2,984	1,525	3,098
Chi	720	985		968	2,015	925	2,130
Dal	781	1,770	968		1,435	520	1,814
LA	2,217	2,984	2,015	1,435		1,895	380
NO	467	1,525	925	520	1,895		2,275
SF	2,595	3,098	2,130	1,814	380	2,275	

Gas Mileage per Rental Car Type

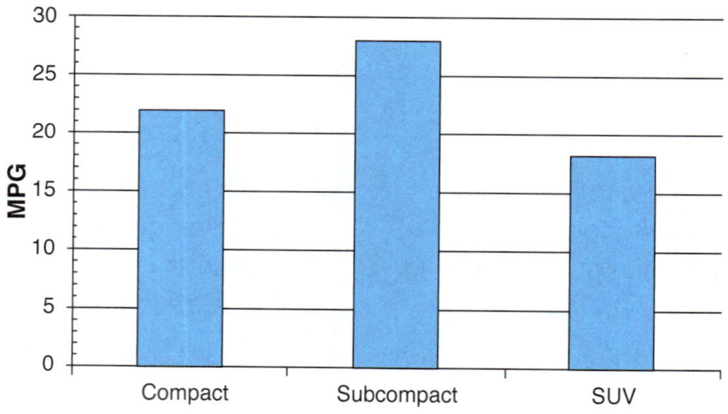

USING TWO DATA SOURCES

Example How much will it cost Greg to drive from Los Angeles to San Francisco if he rents a subcompact car and the price of gasoline is $3.70 per gallon?

Step 1
Gather the information.

The total distance is 380 miles.
A subcompact car gets 28 mpg.

Step 2
Find the number of gallons of gas it will take to get to San Francisco.

380 ÷ 28 ≈ 13.6 gallons

Step 3
Find the total cost of the gasoline.

13.6 × $3.70 = $50.32

It will cost **$50.32** in gas for Greg to drive a subcompact from Los Angeles to San Francisco.

Use the table and graph on the previous page for problems 1 and 2.

1. What will it cost for gas to drive the subcompact car from Boston to Chicago if gas costs $4.05 per gallon?

2. How much more does it cost to drive from Dallas to New Orleans in an SUV instead of a compact car if gas costs $3.50 per gallon?

Use these two graphs for problems 3–6.

2012 Oregon Presidential Election Results

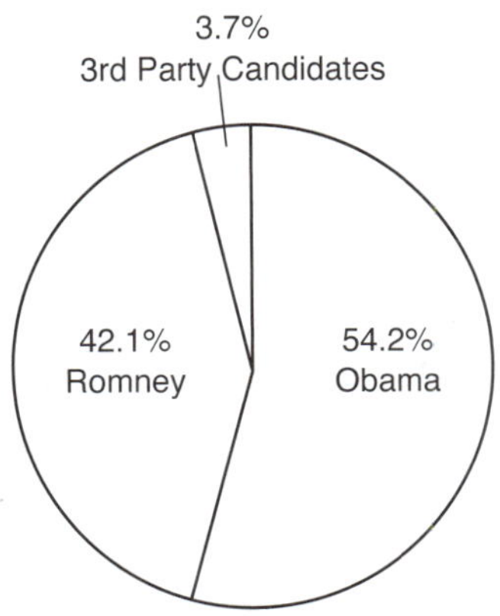

Source: Oregon Elections Division

2012 U.S. Presidential Election Results

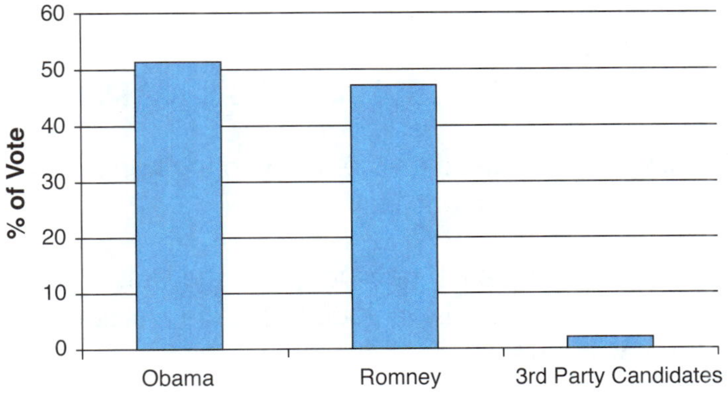

Source: Federal Election Commission

3. If a total of 1,789,270 people voted in Oregon, about how many voted for Obama?

4. Out of a total of 1,789,270 voters, how many more voted for Obama in Oregon than Romney?

5. Did Obama get a larger or smaller percent of the vote in Oregon than in the entire country?

6. How much greater a percentage of the vote did Romney receive nationwide compared to Oregon? Round to the nearest whole percent.

Use these two graphs for problems 7–10.

Average Temperature in Dayton County

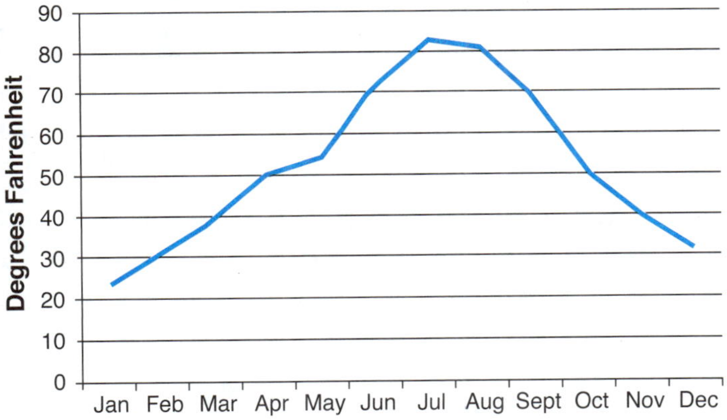

Traffic Accidents in Dayton County

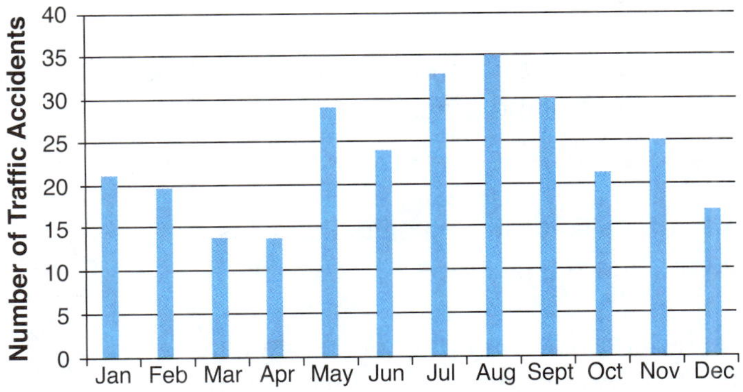

7. In which two months is the temperature usually highest in Dayton County?

8. Does the month with the lowest temperature have the fewest traffic accidents?

9. In which two months are there typically the most traffic accidents in the county?

10. **Explain** What might be some reasons for the slight relationship between temperature and number of accidents in Dayton County?

Answers start on page 167.

Problem Solver

DRAWING CONCLUSIONS

Graphs are often used to prove a position. When you draw a conclusion based on a graph, make sure your conclusion is **justified**. Statements that are justified are supported by facts. Statements that are not justified might be based on exaggerated information or opinions.

Be sure to check the scale of a graph and read axis labels. Some graphs can be misleading and may cause you to draw incorrect conclusions.

READING MISLEADING GRAPHS

Example Based on the graph to the right, is the following statement justified?

"The sale prices of new homes in the United States dropped drastically in 2009 but made a sizeable recovery in 2010."

If you look only at the length of the bars, the conclusion seems justified.

However, if you read the values on the graph, you might draw a different conclusion. Sale prices for new homes fell about $15,000 in 2009—about a 6% decrease.

This graph is misleading because the scale doesn't start at 0. Look at the same graph on a different scale. When the scale starts at 0, you can see that sale prices fell very little in 2009, and they hardly changed at all in 2010.

The statement isn't justified. It isn't supported by the facts of the graph.

Median Price of New Homes Sold in the U.S.

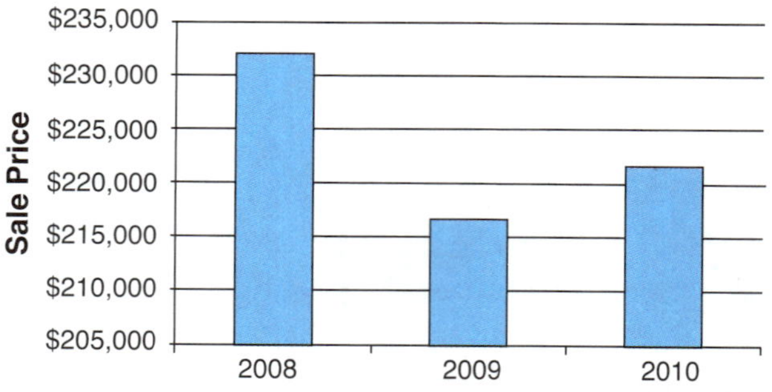

Source: Statistical Abstract of the United States, 2012

Median Price of New Homes Sold in the U.S.

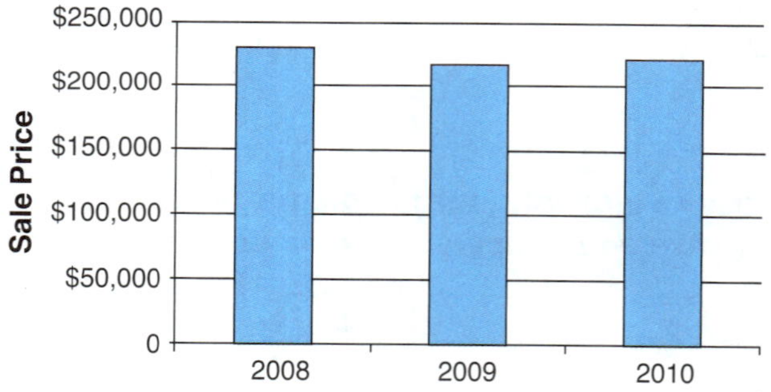

You can also draw unjustified conclusions if you ignore axis labels.

In the graph to the right, you might think the U.S. population has decreased since 1950. However, the vertical axis label states that this graph is measuring the rate of increase. The U.S. population has actually increased the entire time, but it has grown much more slowly since 1950.

U.S. Population Change: 1950 to 2010

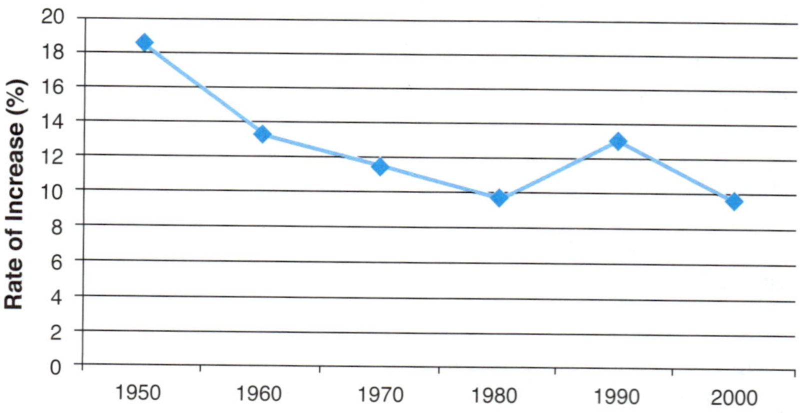

Source: U.S. Census Bureau

Use the graph to determine whether the following statements are justified.

1. Lakers tickets are about 10 times more expensive than Dodgers tickets.

2. More people attend Dodgers games than Lakers games.

3. On average, Lakers tickets are about $65 more than Dodgers tickets.

4. **Explain** What makes this graph misleading? What could be done to make it less misleading?

Use the line graph to determine whether the following statements are justified.

5. The number of women enrolled in colleges has increased considerably since 1970.

6. Women made up a greater percent of college enrollment in 2010 than they did in 1970.

7. The percentage of college students that are women increased less than 5% from 1990 to 2010.

8. **Explain** What makes this graph misleading? What could be done to make it less misleading?

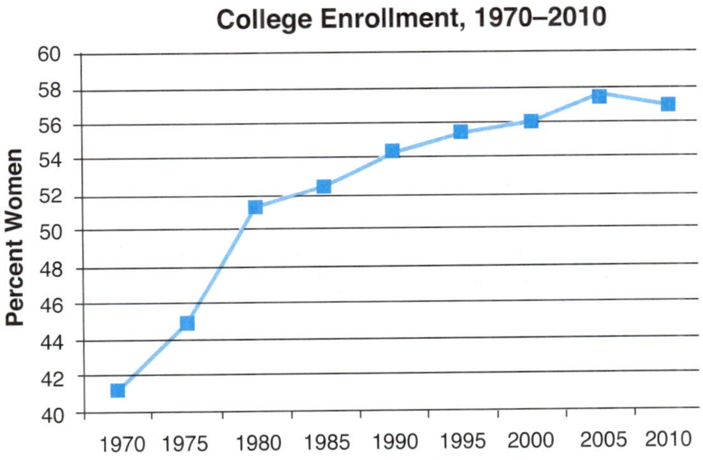

Source: National Center for Education Statistics

Use the bar graph to answer the following questions.

9. Which elementary school's budget decreased the least from Year 1 to Year 2?

10. By about what percent did Franklin's budget decrease?

11. Draw a justifiable conclusion about budget cuts in Hodgson County from Year 1 to Year 2.

12. **Explain** Is this graph misleading? Would you make changes to make it less misleading?

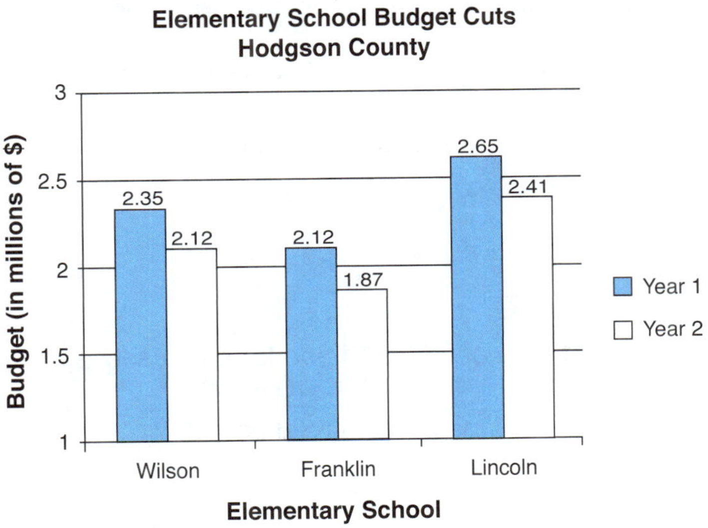

Answers start on page 168.

Unit 1 Review

Use the box plot for problems 1–4.

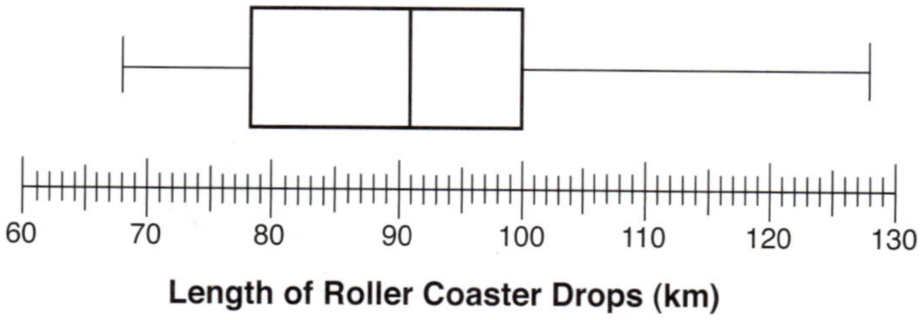

1. What is the range of the data shown on the box plot?

☐

2. What is the median drop length?

☐ km

3. What percent of roller coasters have a drop between 78 km and 100 km?

☐ %

4. What is the upper quartile of the data?

☐ km

Use the histogram for problems 5 and 6.

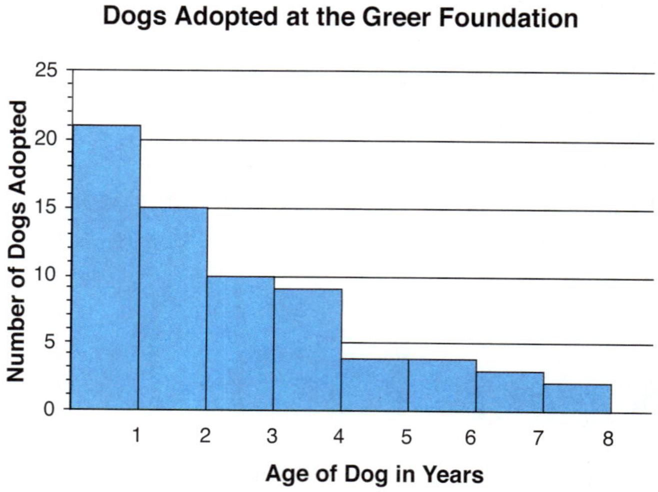

5. Which of the following best describes the histogram?

A. skewed left

B. skewed right

C. symmetric

D. high variation

6. Out of the 68 dogs that were adopted, what percent were 2 years old or older? Round to the nearest percent.

☐ km

Choose which type of correlation the following relationships have.

7. Time spent in the sun and degree of sunburn

- positive correlation
- negative correlation
- no correlation

8. Age at birth of first child and Internet usage

- positive correlation
- negative correlation
- no correlation

9. Hours worked and hours of free time

- positive correlation
- negative correlation
- no correlation

10. A person's age and the amount of gray hair

- positive correlation
- negative correlation
- no correlation

Use the following graph for problems 11 and 12.

Orlando Industries, Results of Advertising Spending

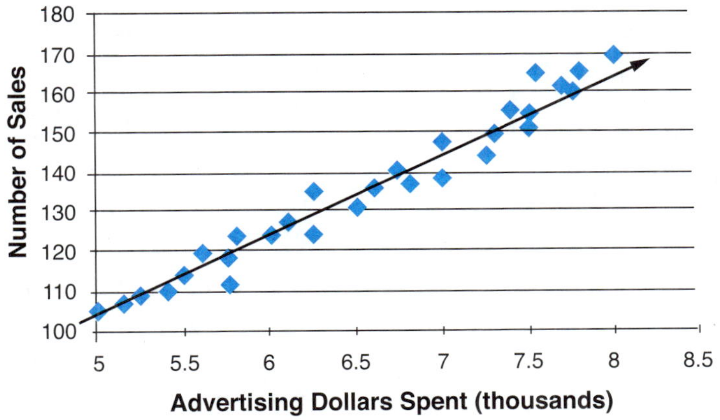

11. In May, Orlando Industries spent $7,000 on advertising and made 150 sales. Where would this value fall?

A. Above the line of best fit

B. Below the line of best fit

C. On the line of best fit

12. If the target sales number is 165, how much should Orlando Industries plan on spending on advertising?

$ ☐

Use the following graphs for problems 13 and 14.

Employees at Shield Manufacturing, by Department

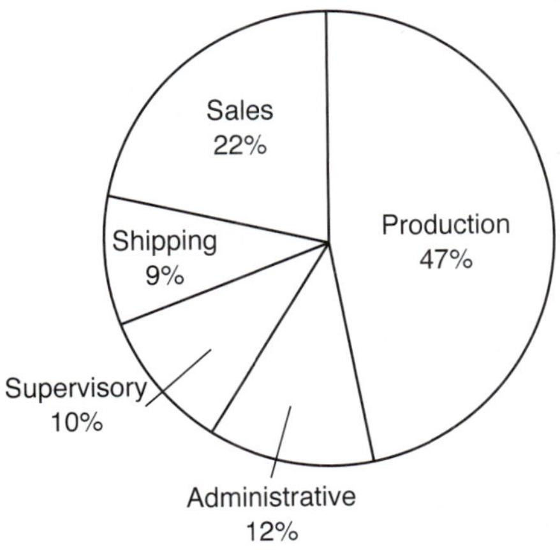

Average Yearly Salary at Shield Manufacturing, by Department

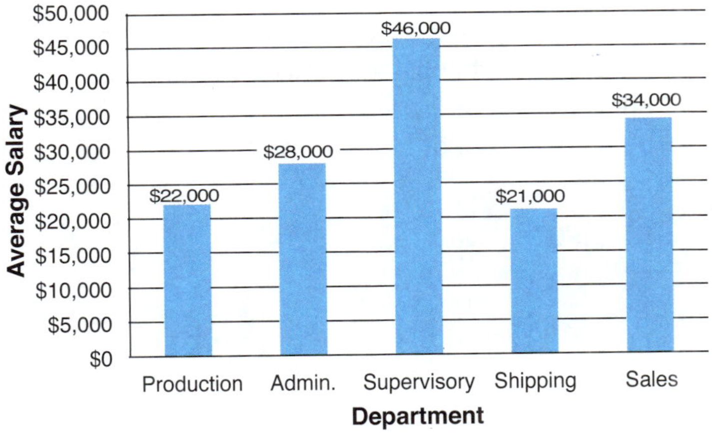

13. If Shield Manufacturing has 200 total employees, how much does the company spend on administrative salaries in one year?

A. $336,000

B. $672,000

C. $1,750,000

D. $5,600,000

14. The production department accounts for the greatest portion of money paid for salaries. Out of 200 employees, which department is paid the next largest portion for salaries?

$ ☐

Solve as directed.

15. A baseball team scores the following number of runs in its first 14 games.

1	2	5	8	4	2	4
4	10	1	3	0	6	5

How many runs must the team score in its 15th game to average 4 runs per game?

A. 3

B. 4

C. 5

D. 6

16. A company's 21 employees averaged 638 travel miles in August. One employee traveled 3,398 miles that month. What was the mean number of travel miles for the other 20 employees?

A. 500

B. 2,760

C. 10,000

D. 13,398

17. A company that manufactures computer chips tests every fifth chip for quality. Out of 3,000 chips tested, 960 were faulty. At that rate, how many chips out of 15,000 would be defective?

A. 320

B. 1,960

C. 4,800

D. 9,000

18. In a class of 25 students, the median test score is 73. Exactly one student scored a 73. How many students scored higher than a 73?

[ ] students

19. A short story contest uses the following scoring system:

Category	Weight
Originality	15%
Creativity	35%
Execution	40%
Mechanics	10%

Paola's story receives a 7 for Originality, a 9 for Creativity, a 6 for Execution, and a 10 for Mechanics. What is her total weighted score?

A. 7

B. 7.2

C. 7.6

D. 8

Use the bar graph for problems 20 and 21.

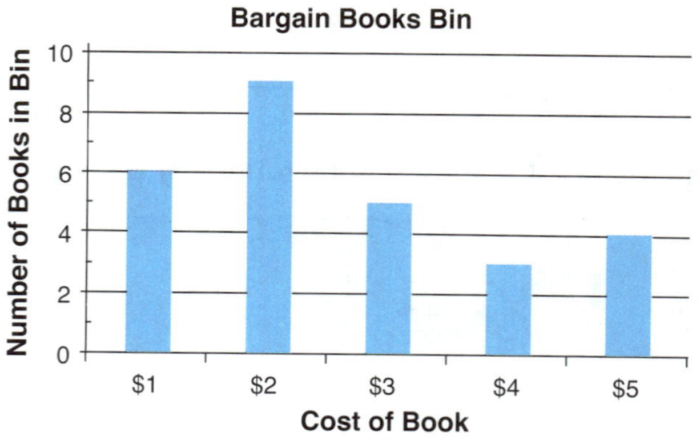

A bargain book bin contains books that range in price from $1 to $5. The bar graph shows the prices of the books in the bin.

20. What price is the mode?

$ [ ]

21. What is the median price?

A. $1

B. $2

C. $3

D. $4

Use the bar graph for problem 22.

22. A newspaper prints the graph to the right and claims that the number of bank robberies in Hastings County has increased dramatically in recent years. Is this conclusion justified? Explain.

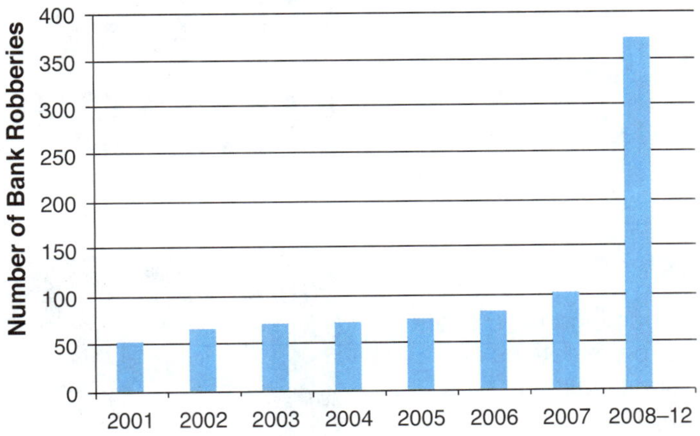

Use the graph and table for problems 23–26.

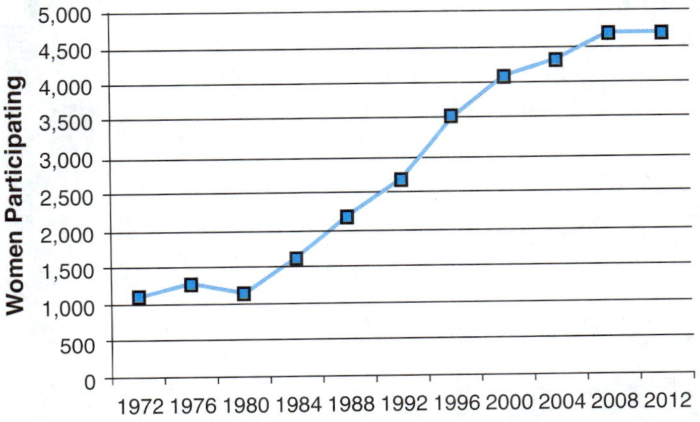

Source: International Olympic Committee

Women's Sport/Discipline	Year Introduced
Shooting, Cycling	1984
Tennis, Table Tennis, Sailing	1988
Judo, Biathlon, Badminton	1992
Soccer, Softball	1996
Weightlifting, Pentathlon, Taekwondo, Triathlon	2000
Wrestling	2004
BMX	2008
Boxing	2012

23. In the year that women's softball was introduced, about how many women participated in the Summer Olympics?

A. 1,996
B. 2,700
C. 3,500
D. 4,000

24. Between which years did the number of women participating in the Summer Olympics increase the most?

A. 1976 and 1980
B. 1980 and 1984
C. 1988 and 1992
D. 1992 and 1996

25. In which year were the most women's sports added to the Summer Olympics?

A. 1992
B. 1996
C. 2000
D. 2004

26. Which of the following describes the current trend in the number of women participating in the summer Olympics?

A. The number has reached 5,000.
B. The number is decreasing.
C. The number is increasing greatly.
D. The number has remained fairly constant.

Answers start on page 168.

UNIT

2

COUNTING AND PROBABILITY

SKILLS

- Counting the number of permutations or combinations
- Finding simple probability
- Finding probability of compound events
- Finding experimental probability

TOOLS AND STRATEGIES

- Making an organized list or tree diagram
- Using probability to make predictions

PROBABILITY is a mathematical way of expressing the chance that something will happen. When you think about how likely it is that a certain event will happen, you are thinking about probability.

To calculate probability, you have to be able to count all the possible ways that something can happen. In this unit, you will learn basic strategies for counting possibilities. You will learn to recognize and count **permutations** and **combinations**.

Very often, the probability of an event is much greater or less than most people would predict. People may take foolish risks or miss opportunities because they underestimate or overestimate the chance of failure or success.

In this unit, you will learn the basics of calculating probability. Understanding these basics will help you make more accurate predictions about the world around you.

COUNTING AND PROBABILITY PROBLEMS

You use counting techniques and probability when you:

- Figure out how many different ways there are to do something
- Estimate how likely it is that something you want will happen
- Play games of chance
- Invest money
- Buy insurance

Read the following items. Describe some of your experiences in writing or discuss them with a partner.

1. Suppose you are planning an outdoor activity, but the weather report says there is a 30% chance of rain. How do you decide whether to keep or change your plans?

2. Have you ever made a friendly bet? Maybe you have entered an office basketball pool just for fun. How do you decide what to bet on? How do you make your decisions?

3. Suppose you are invited to play a new game. How do you decide whether the game is a game of luck, skill, or both? Think of one example of a game of luck and one of a game of skill.

4. Scientists have found that about 10% of people are left-handed. Does that mean that every child has a 10% chance of being left-handed? Explain your thinking.

MATH TALK

You know that when you flip a coin, there is an equal chance that the result will be heads or tails. You have just flipped a coin four times, and the result was heads every time. You are going to flip the coin one more time. Do you think it will come up heads or tails?

Discuss your thinking with a partner.

A friend says that nothing about the coin has changed, so there is still a 50% chance of getting heads. What do you think of this statement?

Problem Solver

COUNTING STRATEGIES

A counting problem asks you to count the possible ways that something could happen. In the rest of this unit, you will learn how to use operations to find the number of possibilities. However, in many simple problems, all you need to do is organize your work and count carefully.

Example The Parkers are taking a photograph of their four children—Alan, Bianca, Callie, and Dennis. They plan to stand in a line. How many different ways can the children line up?

To find the answer, you can use the first letters of their names to represent the children and write all the possible orders.

Step 1
Write the orders that begin with Alan (A).

ABCD
ABDC
Two start with AB.

ACBD
ACDB
Two start with AC.

ADBC
ADCB
Two start with AD.

Step 2
These begin with Bianca (B).

BACD
BADC

BCAD
BCDA

BDAC
BDCA

Step 3
These begin with Callie (C).

CABD
CADB

CBAD
CBDA

CDAB
CDBA

Step 4
These begin with Dennis (D).

DABC
DACB

DBAC
DBCA

DCAB
DCBA

Count the orders. There are **24 possible ways** in all.

If you carefully organize your work, you can often see a pattern that will help you count the possibilities. Once you know that there are 6 ways that start with Alan, you can conclude that there will be 6 ways for each child: $6 \times 4 = 24$ possible ways.

A diagram is another way to organize your work.

Example Emily wants to wear red, white, and blue for a holiday. She can choose from two pairs of sneakers—blue or white. She has three shirts—red, white, or blue—and three pairs of shorts—red, white, or blue. She wants to wear one red item, one white item, and one blue item. How many outfits can she make?

This problem sounds more complicated than it really is. One way to solve the problem is to draw a diagram with columns for each type of clothing. Then draw lines to connect colors to make outfits.

As you can see from the diagram, Emily can make **four possible outfits**.

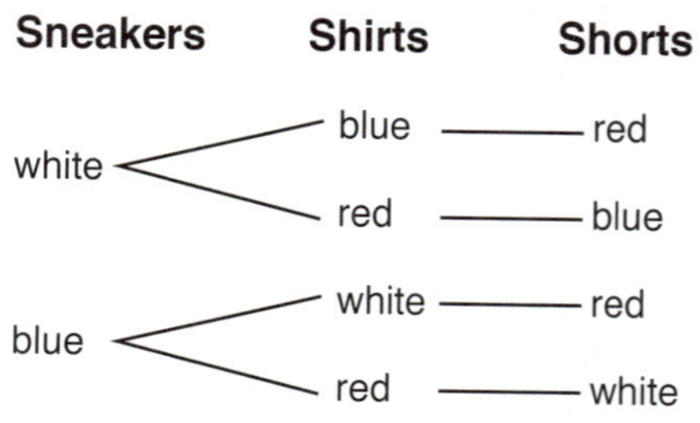

In the problems that follow, try different strategies. You can make an organized list, create a diagram, or even draw a picture. Good problem solvers look for meaningful ways to organize the information in the problem. As they organize, they look for patterns that will help them find the answer.

Solve.

1. A television network is planning its Thursday night lineup. It has 2 sitcoms, 1 drama, and 1 reality show planned for Thursdays. In how many different orders could the programs be shown?

2. A car can seat 2 people in the back and 2 in the front. A mother and her 2 children get in the car. In how many different ways could the family be seated if the mother always has to be the driver and either child can sit up front?

3. Jan is ordering custom shirts for her business. There are 5 colors and 3 styles of shirts to choose from. How many different choices does she have?

4. In the figure, how many different line segments can be drawn that connect exactly two points? (***Note:*** AB is the same as BA.)

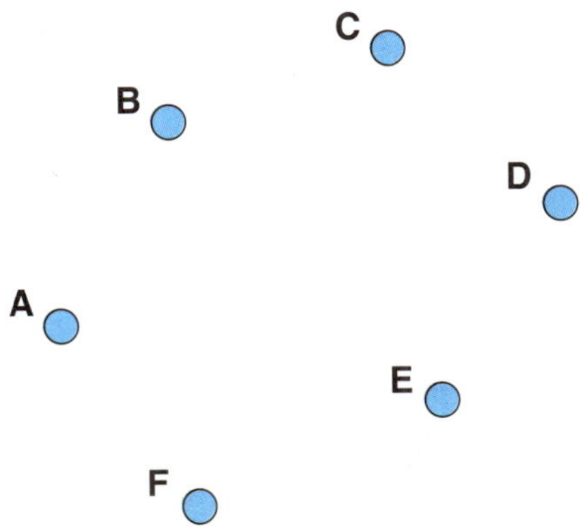

5. The directions for a game say that a player must take two different actions on his or her turn.

The player's choices are:

- Draw a card
- Play a card
- Take a move action
- Capture an opponent
- Buy an item from the bank

The order that the player takes the actions matters. Moving and then playing a card is different from playing a card and then moving.

How many different two-action turns are possible?

6. The truck in the drawing needs to make a delivery to the house in the top right corner. If the truck must stay on the lines and only travel north or east, how many different routes can the truck take to get to the house?

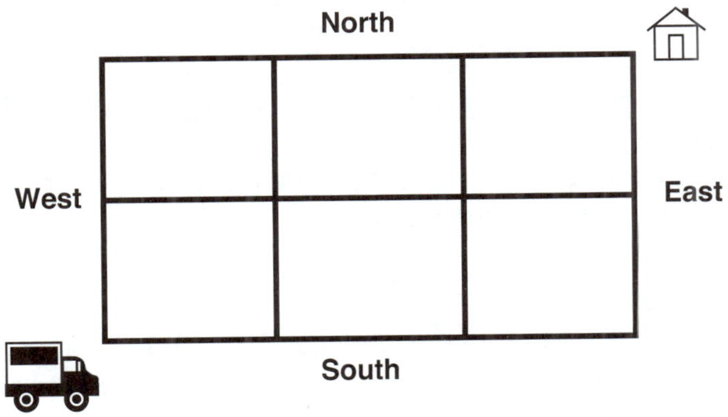

Answers start on page 169.

PERMUTATIONS

In the last lesson, you counted the possible ways that four children could line up for a photograph. A way was considered different if the children were in a different order. When order matters, you are counting the number of **permutations**.

In real life, the word *combination* is often used when really it is a *permutation*. When you use a combination lock, you have to enter the numbers in a certain order. A combination lock should really be called a permutation lock.

You will learn more about combinations in the next lesson. For now, remember that in a permutation problem, the order of the events or items matters.

COUNTING PERMUTATIONS

Example In how many different ways can a student arrange five different books on a shelf?

Step 1
Count the number of choices the student has for each space on the shelf.

There are 5 choices for the first book. After a book is placed in that position, there are 4 choices for the second book. After each book is placed, there is 1 fewer choice. Notice how the choices count down after each book is placed.

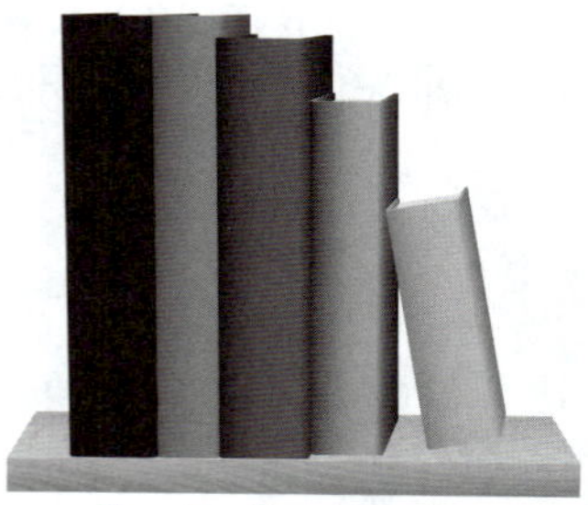

Step 2
Multiply the choices: $5 \times 4 \times 3 \times 2 \times 1 = 120$

There are **120 different ways** the books can be arranged on the shelf.

To solve any permutation problem, simply multiply the choices. Study how this basic idea can be used to solve a more complex problem.

FINDING PERMUTATIONS OF PART OF A GROUP

Example Fifteen runners are in a race. How many different finishes can there be for 1st, 2nd, and 3rd place?

Step 1
Count the number of choices. There are 15 runners who could finish in 1st place, 14 choices for 2nd place, and 13 choices for 3rd place. Remember, once a runner finishes 1st, that runner cannot also finish in 2nd or 3rd place.

Step 2
Multiply the choices: $15 \times 14 \times 13 = 2{,}730$

There are **2,730 different ways** the runners could finish 1st, 2nd, and 3rd.

Will the choices always count down? Not always.

Example A bike combination lock has three dials, each with 10 numbers. To open the lock, you turn the dials. A number can be used *more than once*. How many permutations are possible?

There are 10 choices for each dial: $10 \times 10 \times 10 =$ **1,000**.

Solve.

1. Jasmine has 8 songs in a playlist. The playlist is set to play songs randomly. If she listens to 3 songs, how many different orders are possible if a song may not be repeated?

2. A company assigns a code to each customer. The code has 2 letters followed by a single-digit number from 1 to 9. If the letters must be different, how many codes can the company make? (***Hint:*** There are 26 letters in the alphabet.)

3. For summer reading, Grace must choose 4 books out of a reading list of 12 books. In how many different ways could Grace complete the assignment?

4. Five friends are performing a song for a talent show. They plan to stand in a line. In how many different orders could they stand?

CORE CONNECTIONS: Circular Permutations

When the items in a problem are arranged in a circle, you need an extra step to count the permutations.

Example Four friends sit at a circular table. In how many orders can they sit?

If the friends sat in a line, you could multiply: $4 \times 3 \times 2 \times 1 = 24$ ways.

Since the friends are sitting in a circle, you have to make sure you aren't counting the same order more than once. Look at the tables to the right. The letters A, B, C, and D represent the four friends. What do you notice?

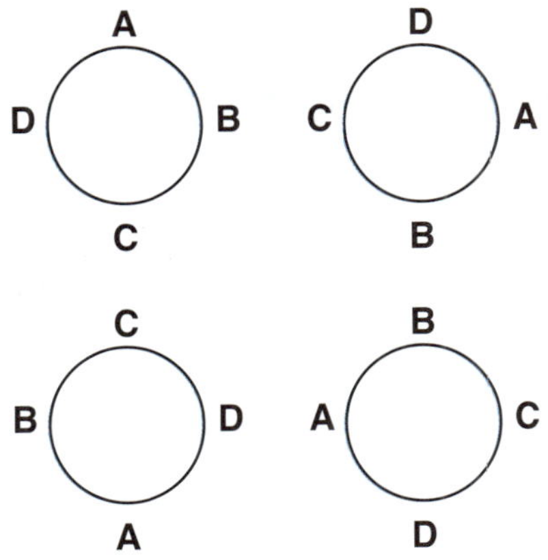

The tables are simply rotated clockwise. Their orders are the same. To count permutations when the items are in a circle, divide by the number of items.

There are 24 ways in a line. There are $24 \div 4 =$ **6 ways in a circle**.

Solve.

1. There are 5 beads on a bracelet. In how many different ways can the beads be arranged?

2. A key ring has 6 keys. In how many different orders can the keys be placed on the ring?

Answers start on page 169.

COMBINATIONS

A **combination** is an arrangement of the items in a set in which the order of the items doesn't matter.

Suppose there are 8 workers in an office. The fire safety team needs 3 people. The manager puts the 8 workers' names on slips of paper and randomly chooses 3 names. Does the order the names are chosen make a difference?

No, the workers will be a group. The order doesn't matter.

This is an example of a combination. Now let's see how to count it.

Leticia		Joshua
Calvin	=	Leticia
Joshua		Calvin

The orders are different, but the group of names is the same.

COUNTING COMBINATIONS

Example Three employees out of a group of 8 will be chosen randomly to be the fire safety team. How many different groupings are possible?

Step 1
Find the permutations for choosing 3 names out of 8.

$8 \times 7 \times 6 = 336$

Step 2
Find the permutations for a group of 3. You are finding how many different ways 3 names can be arranged.

$3 \times 2 \times 1 = 6$

Step 3
Divide.

$336 \div 6 = 56$

There are **56 possible combinations**.

How can you tell the difference between a permutation and a combination problem? You need to use common sense. Remember, if changing the order matters in the problem, it's a permutation.

Decide what kind of problem each situation describes. Write *P* for permutation or *C* for combination.

1. In a lottery, you win a prize if you can guess which 6 balls will be drawn from 30 balls, numbered 1 through 30.

2. A class president and secretary will be randomly chosen from a class of 20 students.

3. A phone company generates new numbers for an area based on the last four digits in a telephone number.

4. Three scoops of different flavors of ice cream are placed in a bowl to make a sundae.

5. A baseball manager chooses 3 players from a group of 15 to bat in the first, second, and third spots in the lineup.

6. Five basketball players are chosen from a group of 12 to start the game.

Solve.

7. From an office with 15 workers, 4 workers are chosen randomly to form a holiday committee. How many different committees could be chosen?

8. Mark has 6 different T-shirts. He puts 2 in a suitcase to take on a trip. How many different groups of 2 T-shirts are possible?

9. A study group of 3 students is formed from a class of 18 students. How many different study groups can be formed?

10. A florist has 6 different kinds of flowers to choose from. If he makes a bouquet using 3 kinds of flowers, how many combinations are possible?

11. A magician tells an audience member to choose 2 cards from a deck of 52 different cards. How many different 2-card pairings are possible?

12. On a reality show, 5 people will be chosen from a group of 30 people to go on to the next round. How many different groups are possible for the next round?

Solve. The problems may involve permutations or combinations.

13. In a cooking contest, teams of chefs will have the chance to cook for a panel of judges. The teams of 3 will be formed randomly from a pool of 20 chefs.

    a. How many different groups of 3 can be formed?

    b. Five teams will take turns competing the first night of the contest. In how many different ways could 5 teams be scheduled?

14. A swim coach chose 4 swimmers for the relay race. He definitely wants Matt to swim the last leg of the race. In how many different orders could the swimmers swim the relay race?

15. In a small office, every employee is assigned a sign-in code. The code has 3 digits. The first digit cannot be a 0. The second digit must be a 5, 6, or 7. The third digit must be different from the first digits. How many 3-digit codes are possible?

16. The blocks below are placed in a row. In how many different orders can you place the blocks if the fully shaded block cannot be placed at either end of the row?

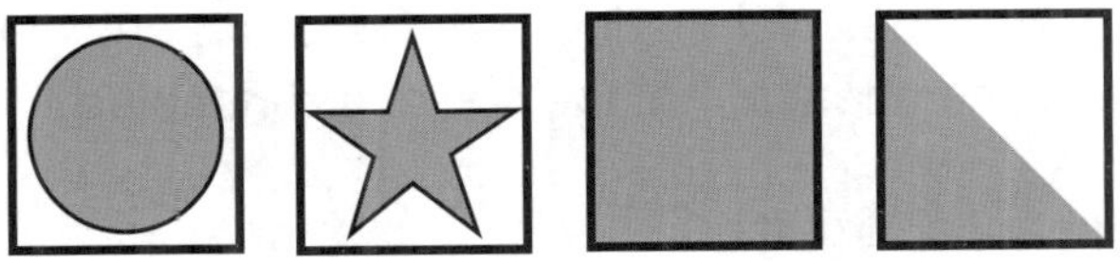

Answers start on page 169.

SIMPLE PROBABILITY

Probability is the likelihood, or chance, of something happening or not happening.

Probability is a ratio. It compares the number of favorable outcomes to the number of possible outcomes. A favorable outcome is the event you are analyzing.

$$\text{probability of an event} = \frac{\text{number of favorable outcomes}}{\text{total number of possible outcomes}}$$

Probability is often written as a percent.

- A probability of 0% means than an event cannot happen.
- A probability of 100% means than an event is certain to happen.

WRITING PROBABILITY AS A NUMBER

Example What is the probability of rolling a 6 using a regular 6-sided die?

Step 1
Count outcomes. Since the die has 6 sides, the total number of possible outcomes is 6. Since there is only 1 side with a 6, the number of favorable outcomes is 1.

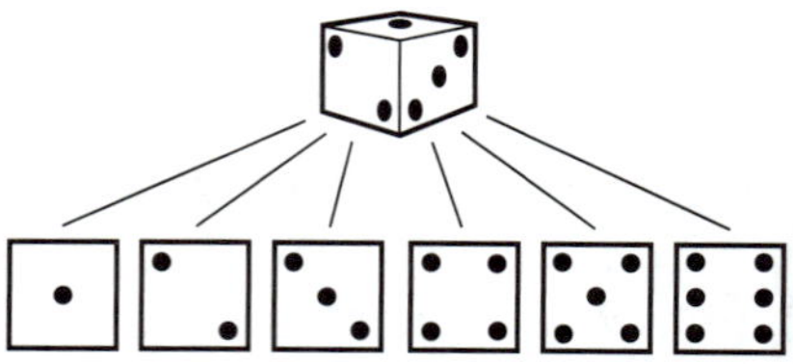

Each face has an equal chance of turning up.

Step 2
Write the ratio.

$$\text{probability of rolling a 6} = \frac{\text{number of favorable outcomes}}{\text{total number of possible outcomes}} = \frac{1}{6}$$

The probability of rolling a 6 is $\frac{1}{6}$ or **1 chance in 6** or **about 17%**.

You can also figure out the probability of something *not* happening.

Example Oscar has 10 pairs of socks. Of the pairs, 3 are black, 2 are gray, 3 are brown, and 2 are blue. Without looking, Oscar reaches into the drawer and draws out a pair of socks. What is the chance that the socks are *not* blue?

The chance of getting blue is $\frac{2}{10}$, which equals $\frac{1}{5}$, or 20%.

To find the chance of *not* getting blue, subtract from 100%. 100% – 20% = **80%**

Finding the **odds** of an event happening is different from finding probability. To find the odds, write a ratio comparing the number of positive outcomes to the number of negative outcomes. The odds of Oscar drawing out a pair of blue socks are 2 blue to 8 non-blue, or **1 to 4.**

Solve as directed. For these problems, write probability as a fraction. Express chance as a percent.

A spinner has 8 equal sections, numbered as shown.

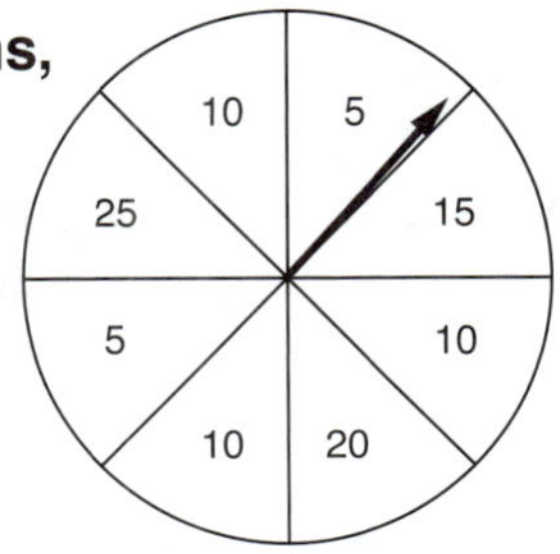

1. If you spin the spinner, what is the probability that you will land on a number greater than 12?

2. What is the chance of the spinner landing on 20 or 25?

3. What is the probability of *not* landing on 10?

A deck has 52 cards in all. There are 4 suits (diamonds, clubs, spades, and hearts). Each suit has 13 cards: an ace, a king, a queen, a jack, and the numbers 2 through 10.

4. If a card is selected at random, what is the chance of choosing a diamond?

5. What is the probability of selecting an ace at random?

6. What is the probability of drawing a card with either a 5 or a 6 of any suit? (***Hint:*** Count the total number of 5s and 6s.)

7. What are the odds of drawing a card from the suit of hearts?

Each question is based on rolling a regular 6-sided die.

8. What is the chance of rolling an even number?

9. What are the odds of rolling a number divisible by 3?

10. What is the probability of *not* rolling a 1?

11. Candice and Lior work in the same office. Two people will be randomly selected to attend a conference. If there are 10 people working in the office, what is the probability that both Candice and Lior will be chosen? (***Hint:*** Use your work with combinations to find the total ways 2 people can be chosen out of 10. Then count the combinations that include Candice and Lior. Use the information to write the probability ratio.)

12. **Explain** Kayla went to the Pizza Pie Café for lunch and ordered the 2-topping special. She made this comment to a friend: "If you order a 2-topping pizza and the toppings are chosen at random, you will have a 1 out of 28 chance of getting a pepperoni and pineapple pizza."

    a. Assume Kayla is correct. How many different toppings are available at the Pizza Pie Café?

    b. How do you know your answer makes sense? Explain your thinking.

Answers start on page 170.

COMPOUND PROBABILITY

Compound probability is the likelihood of several events happening at once or one after the other. Compound probability is found by *multiplying* the probabilities of individual events.

Compound events can be **independent** or **dependent**. Two events are independent when one event doesn't impact the other one. Examples of independent events are flipping a coin more than once or rolling more than one die at a time.

FINDING THE PROBABILITY OF INDEPENDENT EVENTS

Example If you roll two dice, what is the probability of rolling two 6s?

Step 1
Find the probability of the first event.

The probability of rolling a 6 on one of the dice is $\frac{1}{6}$.

Step 2
Find the probability of the second event.

The probability of rolling a 6 on the other die is also $\frac{1}{6}$.

Step 3
Multiply the probabilities.

$$\frac{1}{6} \times \frac{1}{6} = \frac{1}{36}$$

The probability of getting two 6s is $\frac{1}{36}$, or **1 out of 36.**

Two events are dependent when the first event affects the second event.

FINDING THE PROBABILITY OF DEPENDENT EVENTS

Example There are 10 marbles in a bag: 7 are red and 3 are blue. Lisa draws a marble out of the bag. Without replacing the first marble, she draws a second marble from the bag. What is the chance that both marbles are blue?

Step 1
Find the probability of the first event.

$$\frac{3 \text{ blue}}{10 \text{ marbles}} = \frac{3}{10}$$

Step 2
Assume Lisa has drawn out a blue marble. Now there are 9 marbles left and 2 are blue. Find the probability of the second event.

$$\frac{2 \text{ blue}}{9 \text{ marbles}} = \frac{2}{9}$$

Step 3
Multiply the probabilities.

$$\frac{3}{10} \times \frac{2}{9} = \frac{6}{90} = \frac{1}{15} \approx 7\%$$

The probability of drawing 2 blue marbles is **1 out of 15**, or **about 7%**.

If Lisa replaces the marble before drawing out the second marble, the events would be independent. Do you see why? The number of blue and total marbles would not change. If she replaces the marble, the probability is $\frac{3}{10} \times \frac{3}{10} = \frac{9}{100}$, or 9%.

Some problems can be solved by adding probabilities. Mathematicians sometimes call these **either/or probabilities**.

FINDING AN EITHER/OR PROBABILITY

Example If you roll 2 dice, what is the probability of rolling a 7?

When you roll 1 die, there are 6 outcomes.
When you roll 2 dice, there are 36 outcomes: $6 \times 6 = 36$.

Count the ways to make 7.

There are 2 ways using 1 and 6:

$1 + 6$ and $6 + 1$

The probability is $\frac{2}{36}$.

There are 2 ways using 2 and 5:

$2 + 5$ and $5 + 2$

The probability is $\frac{2}{36}$.

There are 2 ways using 3 and 4:

$3 + 4$ and $4 + 3$

The probability is $\frac{2}{36}$.

+	⚀	⚁	⚂	⚃	⚄	⚅
⚀	2	3	4	5	6	7
⚁	3	4	5	6	7	8
⚂	4	5	6	7	8	9
⚃	5	6	7	8	9	10
⚄	6	7	8	9	10	11
⚅	7	8	9	10	11	12

Depending on what 2 numbers you get, the probability could be $\frac{2}{36}$ **or** $\frac{2}{36}$ **or** $\frac{2}{36}$.

The word *or* tells you to add. $\frac{2}{36} + \frac{2}{36} + \frac{2}{36} = \frac{6}{36} = \frac{1}{6}$

The chart proves the answer. The probability of rolling a 7 is **6 out of 36,** or $\frac{1}{6}$.

Solve.

1. When you flip a coin 3 times, what is the probability of getting 3 heads?

2. What is the chance of rolling two 6-sided dice and getting a 5 on both dice?

3. A magician has a deck of 20 cards numbered from 1 to 20. He asks a volunteer to draw 2 cards. What is the probability that both cards are greater than 10?

4. Suppose you roll one 6-sided die. What is the probability of rolling either a 2 or a 3?

5. Jennifer enrolls in 5 classes: Art, Biology, History, Math 1, and Business Math. If her schedule is assigned randomly, what is the probability that her first 2 classes will be math classes?

6. A 3-person committee will be selected from a group of 6 men and 6 women. If names are drawn at random, what is the probability that all 3 people on the committee will be women?

7. Kim buys 10 raffle tickets. Later she finds out that her husband bought 20 tickets. If there are 300 tickets sold in all, what is the chance that either she or her husband will win the prize?

8. A couple has 3 children. Assume there is a 50% chance of having either a boy or a girl. What is the probability that the couple has 2 boys and 1 girl? (***Hint:*** You could make an organized list of the possibilities.)

Answers start on page 170.

USING PROBABILITY FOR PREDICTIONS

Problem Solver

When given a set of collected data, you can use your knowledge of probability to draw conclusions about what will likely happen.

BASING PROBABILITY ON DATA

Rob's Roofing Company offers area homeowners a free estimate for a new roof. The company tracks which zip codes yield positive responses to the special offer. The information is shown on the graph to the right.

Percent of Positive Responses to Free Offer by Zip Code

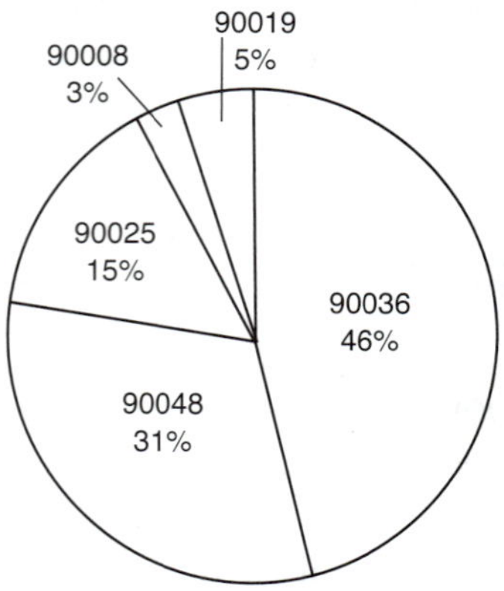

Example 1 From the names of the homeowners who accepted a free estimate, one name is chosen randomly to win a prize. What is the probability that the name will be from the 90036 zip code area?

The graph represents all positive responses to the promotion. The highest percent on the graph is 46%, so there is a 46% chance of drawing a name from zip code 90036.

The probability is **46%**.

Example 2 What is the chance that the name will *not* be from the 90036 zip code?

Subtract: 100% – 46% = 54%
Since 54% is greater than 46%, it is more likely that the name will be drawn from a zip code other than 90036.

The chance is **54%**.

Example 3 If the employees make 220 calls to the 90025 zip code, how many positive responses will they probably get?

You can reasonably conclude that Rob's Roofing Company will continue to have 15% success in the 90025 zip code.

Find 15% of 220.
$220 \times 0.15 = 33$

Of the 220 calls, **33 responses** will probably be positive.

Think about Examples 1 and 2. If someone asked you which zip code the winner would be from, you would probably say 90036. It has the greatest percent. But you would have a better chance of being right if you said the zip code *would not be* 90036.

Solve.

1. During the regular season last year, a basketball player made 448 free throws out of 700. In the current season, he made 35 out of his first 50 free throws.

   What percent of his free throws would you estimate that he will make this season? Explain your thinking.

2. The results of a telephone poll of 500 registered voters about the race for county commissioner are as follows:

Candidates	J. Hinckley	D. Kelly
Female Voters	136	114
Male Voters	154	96
Total	290	210

   a. What percent of the voters polled are supporting J. Hinckley?

   b. Of the next 100 voters polled, how many will probably prefer D. Kelly?

   c. Which candidate will probably receive the greater percent of total support from female voters?

Problems 3 and 4 refer to the following information.

You and a friend are waiting in line near a busy street. Your friend challenges you to make a prediction. "Of the next 100 cars that go by, how many do you think will be blue or white?" You know the following information that might help you make a prediction.

- Last year, 22% of all new cars were white. White is the most popular car color in Europe and North America.
- There are almost 3 times as many white cars as blue cars sold each year.
- Of the last 10 cars that went by on the street, 4 were white. None were blue.

3. What number of white cars would you predict?

4. What number of blue cars would you predict? Explain your thinking.

CORE CONNECTIONS: Area and Probability

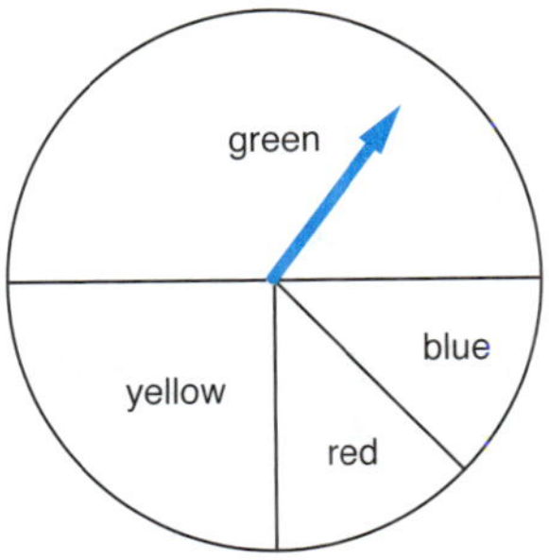

To solve probability problems about area, find out how the area of a certain space compares to the total area.

Think of a spinner as a circle graph. Each section is a certain percent of the whole. The percent is the chance of landing in that section.

Solve.

1. A square target has three sections. Based on the area, what is the percent chance of hitting the shaded squares?

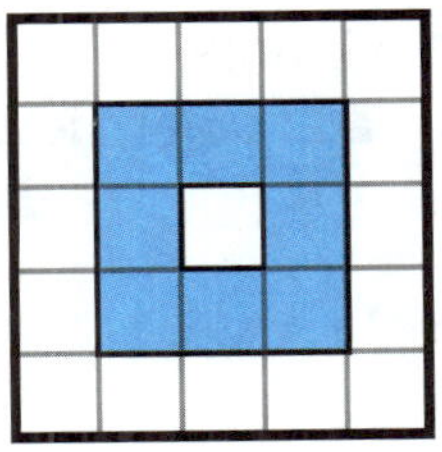

2. On the spinner above, what are the odds of landing on red or blue?

Answers start on page 170.

EXPERIMENTAL PROBABILITY

In real life, you don't always have all the data you would like. Sometimes you have to conduct an experiment to gather data. **Experimental probability** is calculated using the results of an experiment. Each repetition of the experiment is called a **trial**.

Experimental probability is the ratio of favorable outcomes to the number of trials.

$$\text{experimental probability} = \frac{\text{favorable outcomes}}{\text{total number of trials}}$$

USING AN EXPERIMENT TO FIND PROBABILITY

Example 1 A bag of candy has identically shaped pieces of 4 different flavors. Olivia draws a piece of candy from the bag and replaces it. She repeats the experiment 50 times and draws orange 15 times. What is the probability of drawing orange from the bag?

Write a ratio and simplify. $\frac{\text{favorable events}}{\text{total number of trials}} = \frac{15}{50} = \frac{3}{10} = \mathbf{30\%}$

Example 2 If there are 80 pieces of candy in the bag, how many are likely to be orange?

Use the probability from Example 1 to make a prediction.

Find 30% of 80. $80 \times 0.3 = 24$

Based on the experiment, **24** are probably orange.

When the bag is opened, the contents are 20 cherry, 24 grape, 20 orange, and 16 lemon.

The mathematical probability of getting orange is $\frac{20}{80} = \frac{1}{4} = 25\%$. The experiment gave a different result, but it was close enough to make a reasonable prediction.

Read about the experiment. Then answer the questions.

James is testing a new game. He completes 20 turns and records the points the player would earn (see table).

7	3	10	4	10
10	5	12	9	13
8	9	11	8	11
7	7	18	6	10

1. Based on the experiment, what is the chance of earning 10 or more points on 1 turn?

2. In 50 turns, how many times would a player expect to earn less than 5 points?

You may need to use logical reasoning to solve an experimental probability problem.

Example Bag A and Bag B each contain 20 marbles, some blue and some white. You know that one of the bags is 50% blue marbles and the other is 75% blue marbles.

During an experiment, a marble is drawn from a bag and then replaced. The experiment is repeated 10 times. Which bag has 50% blue marbles?

Bag A

white	blue
white	blue
white	white
white	blue
white	white

Bag B

white	blue
blue	blue
blue	blue
blue	white
blue	blue

Consider the results. Bag A has 3 blue and 7 white. Bag B has 8 blue and 2 white. You might expect the bag with 50% blue to have 5 white and 5 blue, but experiments don't always turn out exactly as you would expect.

The other bag is 75% blue. Bag B has many more blue than white. You can conclude that **Bag B is 75% blue**, which means that **Bag A is 50% blue**.

Solve using the given data.

Each of the spinners is spun 8 times and the numbers are recorded.

Experiment #1: 2, 1, 4, 3, 3, 3, 4, 3

Experiment #2: 2, 4, 4, 4, 3, 4, 4, 4

Experiment #3: 3, 2, 2, 2, 4, 3, 4, 4

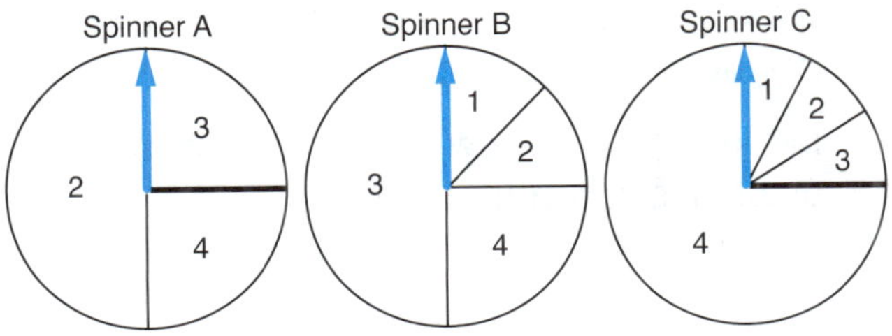

3. Which spinner was used in Experiment #1?

4. Which spinner was used in Experiment #2?

5. Which spinner was used in Experiment #3?

6. **Explain** Suppose you are playing a game where the goal is to get the lowest score. Which spinner would you choose? Explain your reasoning.

A card game has 10 blue cards (B), 10 red cards (R), and 10 white cards (W). The cards are shuffled and 3 cards are dealt. Then the cards are replaced and the deck is again shuffled. This experiment is repeated 20 times. The results of the 20 trials are shown in the table below.

Trial	1	2	3	4	5	6	7	8	9	10	11	12	13	14	15	16	17	18	19	20
Card 1	R	R	B	R	W	B	B	R	W	W	B	R	W	B	R	W	B	R	R	B
Card 2	W	B	R	W	W	B	B	W	B	R	W	B	W	R	B	B	R	B	W	W
Card 3	W	B	R	R	W	B	R	W	R	B	W	R	W	W	R	W	R	R	B	B

7. What is the experimental probability of getting 3 cards of the same color?

8. Based on the experiment, what is the chance of getting 3 different colors?

9. **Explain** Using the data, Doug claims there is a 0% chance of getting 3 red cards. Use your knowledge of mathematical probability to prove Doug is wrong. Why do you think the experiment produced a false conclusion?

Answers start on page 171.

Unit 2 Review

Solve the following problems.

1. A cupcake company offers 4 kinds of cake, 8 kinds of frosting, and 5 kinds of sprinkles. If each cupcake has 1 from each category, how many different cupcakes can the company make?

[ ] cupcakes

2. A diagonal is a line segment that connects two non-adjacent vertices. In the drawing, a segment drawn from *A* to *C* is a diagonal. *AB* is not a diagonal.

In the drawing below, how many different diagonals can you draw?

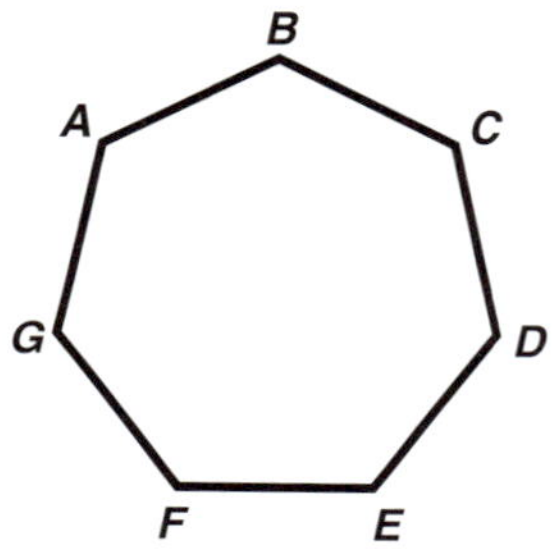

[ ] diagonals

3. There are two stacks of cards. Stack A has five cards, numbered 4, 9, 3, 8, and 6. Stack B has four cards, numbered 1, 5, 7, and 4. One card is drawn at random from each stack. What is the chance that both cards have the same number?

A. 5%

B. 8%

C. 9%

D. 16%

4. Nine athletes are running in a race. Ribbons will be awarded to 1st, 2nd, 3rd, and 4th places. How many different finishes are possible?

[ ] finishes

Use the information to answer problems 5–7.

A bag contains jelly beans: 8 purple, 12 red, 6 pink, and 4 black. You select 1 jelly bean at random.

5. What is the chance that the jelly bean is pink?

[ ] %

6. What are the *odds* that you will get red?

A. 2 to 3

B. 2 to 5

C. 1 to 11

D. 1 to 12

7. What is the probability of *not* getting pink or black?

A. $\frac{2}{5}$

B. $\frac{1}{2}$

C. $\frac{2}{3}$

D. $\frac{3}{4}$

8. A volleyball team has 12 members. For the team photo, 6 will kneel in the front row. How many groups of 6 players could be chosen to kneel in the front?

[ ] groups

9. The letters D, I, V, I, D, and E are written on identical cards and one is chosen at random. Which of the following events has the *greatest* chance of happening?

A. Choosing V or E

B. Not choosing D or I

C. Choosing D or I

D. Not choosing I or E

10. Two regular 6-sided dice are rolled. What is the probability that either exactly one of the die is a 5 or that the total of the two dice is 3?

A. $\frac{1}{6}$

B. $\frac{1}{5}$

C. $\frac{1}{4}$

D. $\frac{1}{3}$

Use the information to answer problems 11 and 12.

In the first 25 games of May, the Jammers won 12 games and lost 13. The number of runs that the Jammers scored in these games is shown below. A shaded box indicates that the team won the game.

9	4	1	**6**	0	3	4
4	0	1	**7**	6	**2**	**6**
3	**8**	**6**	3	**9**	3	**4**
9	2	4	**7**			

11. If the team scores exactly 4 runs, what is the percent chance that it will win that game?

A. 50%

B. 40%

C. 20%

D. 10%

12. There are 162 games in a baseball season. Based on the information in the table, in how many games during the season will the Jammers score at least 6 runs?

A. 16

B. 40

C. 65

D. 78

Use the information to answer problems 13 and 14.

On April 10, the probability that it will rain in Cedar Break is 3 out of 5, or 60%. The probability that it will rain in Shadow Falls is 2 out of 5, or 40%.

13. What is the chance that it will rain in Cedar Break but not in Shadow Falls?

A. 20%

B. 25%

C. 36%

D. 50%

14. Which of the following events would have the greatest probability of happening?

Event A	**Event B**
It will rain in exactly one town.	It will either rain in both towns or in neither town.

Choose one: Event A
Event B

Explain your reasoning in the box below.

Answers start on page 171.

UNIT 3

SYSTEMS OF EQUATIONS AND INEQUALITIES

SKILLS

- Solving systems of equations using substitution and elimination
- Using systems to solve word problems
- Solving linear inequalities
- Graphing linear inequalities

TOOLS AND STRATEGIES

- Using graphs to solve systems of equations
- Interpreting graphs of systems of equations and inequalities

A SYSTEM is a group of related elements that work together for a common goal. An airport has a system of runways, as well as an organized network of computers and air traffic controllers to keep travelers safe. The human body has many systems—the circulatory system moves blood throughout the body, while the digestive system takes in food and nutrients and eliminates waste.

In mathematics, equations and inequalities form systems. Systems have more than one variable. In this unit, you will learn how to write and solve systems with two variables: x and y.

The lessons in the unit will present multiple ways of solving systems. Take the time to understand each method. Each way becomes another tool in your algebra toolbox. Good test takers know that systems are the key to solving complex problems simply.

USING SYSTEMS TO SOLVE PROBLEMS

Systems can be used:

- By businesses to maximize profits
- In sports to create complex schedules
- By hospitals to minimize the costs of caring for patients
- By computer programmers to develop applications

Read the following items. Describe some of your experiences in writing or discuss them with a partner.

1. Think of a formula that has at least two or more variables. What do the variables represent? How do the variables work together to make a statement?

2. Have you ever wanted to open your own business? What kind of business would you choose? What are some of the factors you would need to consider to make sure your business is successful? Variables could represent these factors to form a system.

3. Think about the nervous system in the human body. What parts of the body are in the nervous system? You may want to use another system—the Internet—to research this. What is the purpose of the nervous system?

MATH TALK

Systems express relationships between numbers.

Think of two numbers. Write a number puzzle about the numbers.

For example, you might write:

The sum of two numbers is 20.
One number is three times the other number.
What are the numbers?

Trade puzzles with a partner. Can you use the clues to figure out the numbers? If so, you have just solved a system.

(The numbers in the example are 15 and 5.)

SYSTEMS OF EQUATIONS

A **system of equations** is a collection of equations that use the same variables. In a system, a variable represents the same quantity in each equation.

Suppose the area of a rectangle is 40 square inches and the perimeter of the same rectangle is 28 inches. Area and perimeter are found using length (L) and width (W).

You can use this information to write a system about the rectangle. Both equations use the same variables, and both are about the same rectangle. The brace shows the equations are part of a system.

brace → $\begin{cases} L \cdot W = 40 \\ 2L + 2W = 28 \end{cases}$

The values for L and W that make the first equation true will also make the second equation true. This system is true when $L = 10$ and $W = 4$.

$$\begin{cases} 10 \cdot 4 = 40 \\ 2(10) + 2(4) = 28 \end{cases}$$

Solving a system means finding the values of the variables. One way to solve a system is by graphing the equations.

SOLVING A SYSTEM BY GRAPHING

Example Solve: $\begin{cases} y = -x + 3 \\ y = \frac{1}{2}x - 3 \end{cases}$

TIP

See Math Sense 2: Focus on Problem Solving, *Unit 5, for a complete review of graphing a line.*

Step 1

Graph both equations on the same grid.

Remember, you can always graph a line by making a table of values and then plotting points.

$y = -x + 3$

x	y
−1	4
0	3
1	2

$y = \frac{1}{2}x - 3$

x	y
−2	−4
0	−3
2	−2

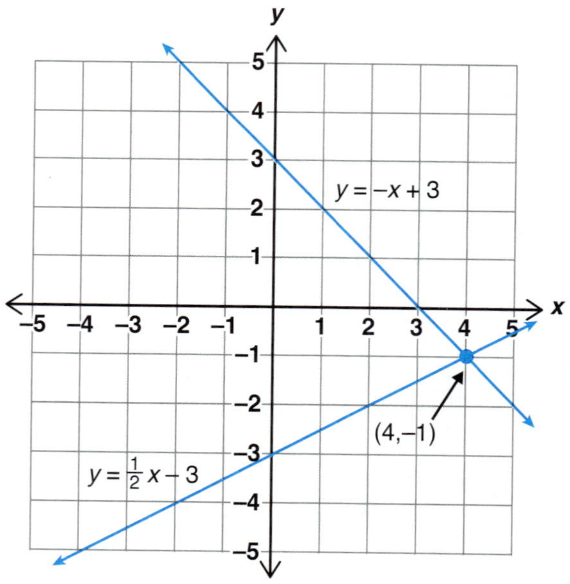

Step 2

The lines intersect at point (4,−1). The x- and y-coordinates of this point are the solution to the system: $\boldsymbol{x = 4}$ and $\boldsymbol{y = -1}$.

Step 3

Check your answer. Substitute the values into the system. Make sure both equations are true.

$y = -x + 3$
$-1 = -(4) + 3$
$-1 = -1$ True

$y = \frac{1}{2}x - 3$
$-1 = \frac{1}{2}(4) - 3$
$-1 = 2 - 3$
$-1 = -1$ True

Most systems have one solution that is written as an ordered pair. In special situations, a system may have no solution or an infinite number of solutions.

On a graph, a system with no solution has parallel lines that never intersect.

A system with infinite solutions appears to be one line because both lines occupy the same space. Every point on one line coincides with a point on the other line.

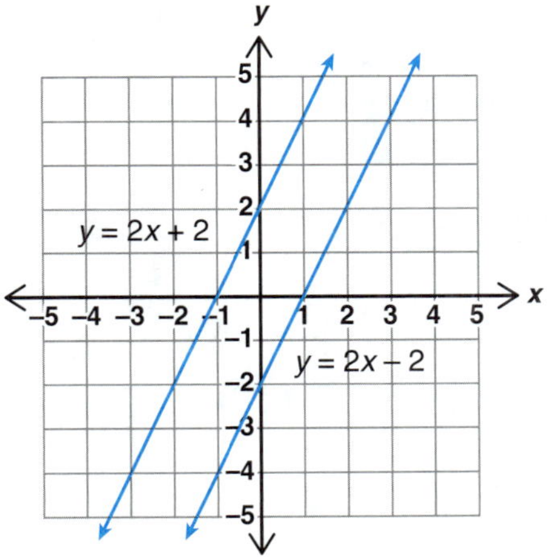

No solution

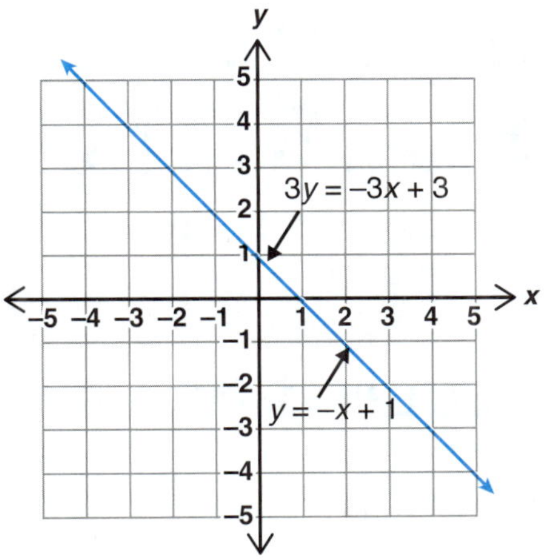

Infinite solutions

Solve by graphing. Write the solution as an ordered pair or write *none* or *infinite*.

1.

$$\begin{cases} y = -x - 4 \\ y = 2x - 1 \end{cases}$$

Solution: ____________________

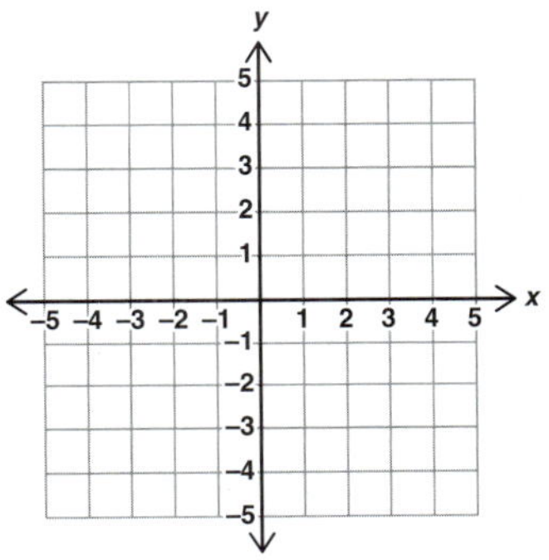

4.

$$\begin{cases} y = x - 3 \\ y = -\frac{3}{4}x + 4 \end{cases}$$

Solution: ____________________

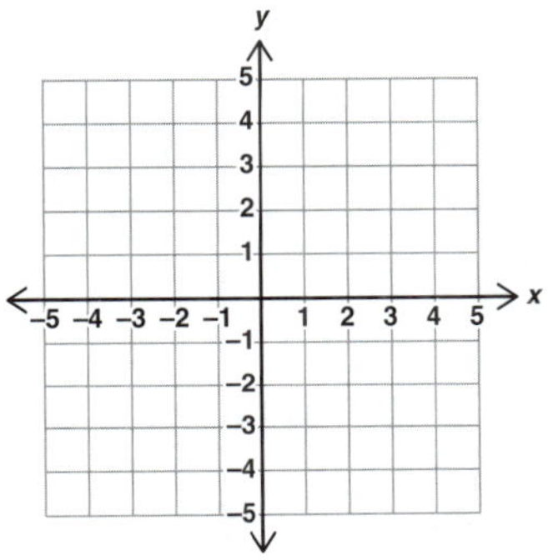

2.

$$\begin{cases} y = -x + 1 \\ y = -6x - 4 \end{cases}$$

Solution: ____________________

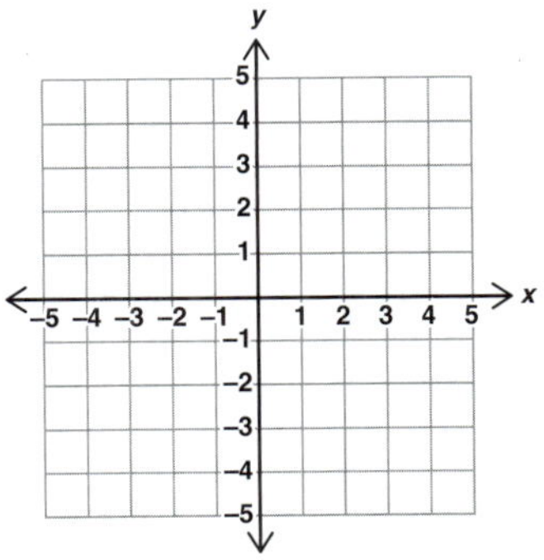

5.

$$\begin{cases} y = 2x \\ 3y = 6x \end{cases}$$

Solution: ____________________

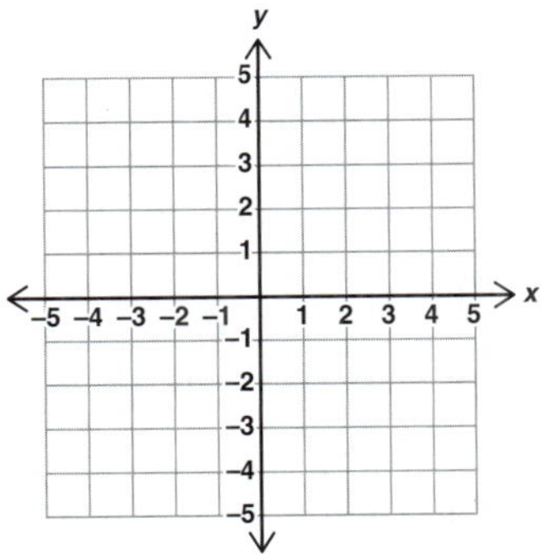

3.

$$\begin{cases} y = 2x - 4 \\ y = 2x - 3 \end{cases}$$

Solution: ____________________

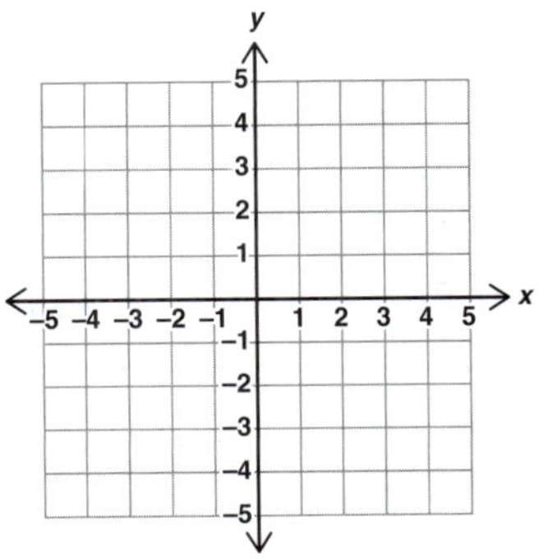

6.

$$\begin{cases} y = -\frac{1}{3}x - 3 \\ y = -2x + 2 \end{cases}$$

Solution: ____________________

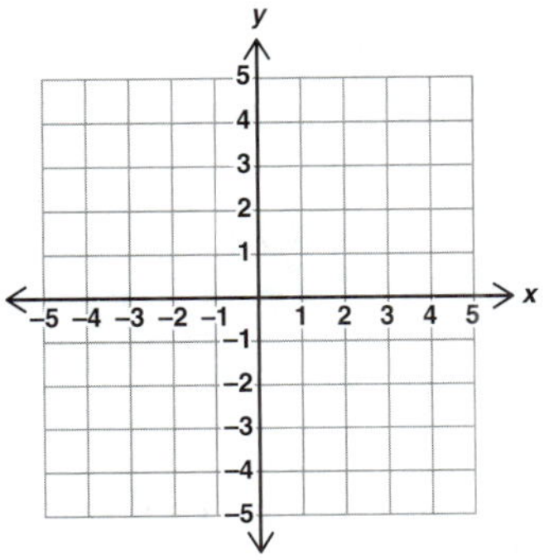

Answers start on page 172.

SOLVE SYSTEMS USING SUBSTITUTION

The **substitution method** for solving a system substitutes an expression from one equation into the other.

SOLVING A SYSTEM USING SUBSTITUTION

Example Solve: $\begin{cases} 3x - y = -2 \\ y = 18 - 5x \end{cases}$

Look at the second equation: $y = 18 - 5x$. It states that the variable y is equal to the expression $18 - 5x$. You can substitute this expression for y in the first equation.

Step 1
Substitute $18 - 5x$ for y in the first equation.

$$\begin{aligned} 3x - y &= -2 \\ 3x - (18 - 5x) &= -2 \end{aligned}$$

Step 2
Solve for x.

$$\begin{aligned} 3x - (18 - 5x) &= -2 \\ 3x - 18 + 5x &= -2 \\ 8x - 18 &= -2 \\ 8x &= 16 \\ x &= 2 \end{aligned}$$

Step 3
Substitute the value for x in the second equation and solve for y.

$$\begin{aligned} y &= 18 - 5x \\ y &= 18 - 5(2) \\ y &= 18 - 10 \\ y &= 8 \end{aligned}$$

Write the solution as an ordered pair. The solution is **(2, 8)**.

Check: Make sure the solution works in both equations in the system.

$$\begin{aligned} 3x - y &= -2 \\ 3(2) - (8) &= -2 \\ 6 - 8 &= -2 \\ -2 &= -2 \quad \text{True} \end{aligned}$$

$$\begin{aligned} y &= 18 - 5x \\ 8 &= 18 - 5(2) \\ 8 &= 18 - 10 \\ 8 &= 8 \quad \text{True} \end{aligned}$$

The solution is correct since it makes both equations true.

Sometimes you will need to rewrite one of the equations to isolate a variable. You can isolate either x or y.

Example Solve: $\begin{cases} x - 2y = -2 \\ 8x - 7y = -16 \end{cases}$

Look for a variable that would be easy to isolate. The first equation has an x without a number next to it. Isolate the x.

$$\begin{aligned} x - 2y &= -2 \\ x &= 2y - 2 \end{aligned}$$

Step 1
Substitute $2y - 2$ for x in the second equation.

$$\begin{aligned} 8x - 7y &= -16 \\ 8(2y - 2) - 7y &= -16 \end{aligned}$$

Step 2
Solve for y.

$$\begin{aligned} 8(2y - 2) - 7y &= -16 \\ 16y - 16 - 7y &= -16 \\ 9y - 16 &= -16 \\ 9y &= 0 \\ y &= 0 \end{aligned}$$

Step 3
Substitute 0 for y in the first equation and solve for x.

$$\begin{aligned} x - 2y &= -2 \\ x - 2(0) &= -2 \\ x - 0 &= -2 \\ x &= -2 \end{aligned}$$

The solution is **(–2, 0)**.

It's very important that you use both equations when you are solving a system. If you isolate the variable in the *first* equation, substitute the expression into the *second*. Then put the solution from that step back into the *first* equation. The equations are related, so you need to involve both to find the answer.

Solve. Write the solution as an ordered pair (x, y).

1. $\begin{cases} y = 2x + 4 \\ 7x - y = -4 \end{cases}$

   (***Hint:*** Substitute $2x + 4$ for y in the second equation.)

2. $\begin{cases} 7x - 6y = -15 \\ y = -6x - 19 \end{cases}$

3. $\begin{cases} y = -2x \\ -8x - y = 12 \end{cases}$

4. $\begin{cases} y = 6x + 18 \\ -2x + 7y = 6 \end{cases}$

5. $\begin{cases} -7y + 8x = -14 \\ x = 5y - 10 \end{cases}$

6. $\begin{cases} x + 4y = 16 \\ -6x - 7y = 6 \end{cases}$

7. $\begin{cases} x - 2y = 19 \\ -4x - 2y = 4 \end{cases}$

8. $\begin{cases} -3x + y = -10 \\ -7x + 6y = -16 \end{cases}$

9. $\begin{cases} -8x + y = 8 \\ 6x + 5y = -6 \end{cases}$

10. $\begin{cases} 4x - y = -23 \\ x - 6y = 0 \end{cases}$

Answers start on page 172.

SOLVE SYSTEMS USING ELIMINATION

Another method for solving systems involves eliminating a variable by combining the equations.

SOLVING A SYSTEM USING ELIMINATION

Example Solve: $\begin{cases} -7x - 9y = 17 \\ 3x + 9y = 3 \end{cases}$

Step 1
Add the equations to eliminate the y-variable.

$$\begin{array}{r} -7x - 9y = 17 \\ 3x + 9y = 3 \\ \hline -4x + 0 = 20 \end{array}$$

Step 2
Solve for x.

$$\begin{aligned} -4x &= 20 \\ x &= -5 \end{aligned}$$

Step 3
Substitute –5 for x in one of the equations and solve for y.

$$\begin{aligned} -7x - 9y &= 17 \\ -7(-5) - 9y &= 17 \\ 35 - 9y &= 17 \\ -9y &= -18 \\ y &= 2 \end{aligned}$$

The solution is **(–5, 2)**.

Check the solution in the other equation.

$$\begin{aligned} 3x + 9y &= 3 \\ 3(-5) + 9(2) &= 3 \\ -15 + 18 &= 3 \\ 3 &= 3 \quad \text{True} \end{aligned}$$

You may need to multiply one or more of the equations before combining. Remember, your goal is to get two terms that, when added, will equal zero.

MULTIPLYING BEFORE COMBINING EQUATIONS

Example Solve: $\begin{cases} -5x + 6y = -8 \\ 9x - 4y = 28 \end{cases}$

Step 1
Choose which variable to eliminate. This example multiplies both equations so the y-coefficients equal 12 and –12.

Step 2
Multiply every term in the first equation by 2.

$$\begin{aligned} -5x + 6y &= -8 \\ 2(-5x + 6y) &= 2(-8) \\ -10x + 12y &= -16 \end{aligned}$$

Multiply every term in the second equation by 3.

$$\begin{aligned} 9x - 4y &= 28 \\ 3(9x - 4y) &= 3(28) \\ 27x - 12y &= 84 \end{aligned}$$

Step 3
Add the resulting equations and solve for x.

$$\begin{array}{r} -10x + 12y = -16 \\ 27x - 12y = 84 \\ \hline 17x + 0 = 68 \end{array}$$

$$\begin{aligned} 17x &= 68 \\ x &= 4 \end{aligned}$$

Step 4
Substitute 4 for x in one of the original equations and solve for y.

$$\begin{aligned} -5x + 6y &= -8 \\ -5(4) + 6y &= -8 \\ -20 + 6y &= -8 \\ 6y &= 12 \\ y &= 2 \end{aligned}$$

The solution is **(4, 2)**. Check the solution in the other equation.

$$\begin{aligned} 9x - 4y &= 28 \\ 9(4) - 4(2) &= 28 \\ 36 - 8 &= 28 \\ 28 &= 28 \quad \text{True} \end{aligned}$$

Good test takers use the problem-solving method that seems best for the situation. Use substitution when one of the variables will be easy to isolate. Use elimination when each variable has a number before it.

Solve. Write the solution as an ordered pair (x, y).

1. $\begin{cases} 9x - 6y = 0 \\ 7x - 3y = 10 \end{cases}$

(***Hint:*** Multiply every term in the second equation by –2.)

2. $\begin{cases} -3x - 4y = -1 \\ x + 3y = -3 \end{cases}$

3. $\begin{cases} 6x + 8y = 10 \\ -2x - 9y = 22 \end{cases}$

4. $\begin{cases} 7x + 22y = -28 \\ 3x - 11y = -12 \end{cases}$

5. $\begin{cases} -2x + y = 3 \\ 3x - 7y = 12 \end{cases}$

6. $\begin{cases} -9x + 4y = 12 \\ -11x + 6y = 8 \end{cases}$

7. $\begin{cases} -7x - 2y = 6 \\ 6x + 5y = 8 \end{cases}$

8. $\begin{cases} 3x - 4y = 15 \\ 10x - 9y = 11 \end{cases}$

9. $\begin{cases} -2x + 3y = -9 \\ 5x - 4y = 33 \end{cases}$

10. $\begin{cases} 9x - 2y = -22 \\ 7x - 5y = -24 \end{cases}$

Answers start on page 173.

Problem Solver

WRITE SYSTEMS TO SOLVE WORD PROBLEMS

Systems are very helpful in solving word problems because you can use more than one variable to represent the information in the problem.

Example The sum of two numbers is 15. Nine times the first number is 3 more than twice the second. What are the two numbers?

Step 1
Write a system to represent the information in the problem. You can use x and y for the two numbers.

Written using words	Written using algebra
The sum of two numbers is 15.	$x + y = 15$
Nine times the first number is 3 more than twice the second.	$9x = 3 + 2y$

Write the system: $\begin{cases} x + y = 15 \\ 9x = 3 + 2y \end{cases}$

Step 2
Choose a method for solving the system. It will be easy to isolate a variable in the first equation, so substitution is the best method.

Step 3
Solve.

Isolate a variable.

$$x + y = 15$$
$$y = 15 - x$$

Substitute $15 - x$ for y in the other equation.

$$9x = 3 + 2y$$
$$9x = 3 + 2(15 - x)$$
$$9x = 3 + 30 - 2x$$
$$9x = 33 - 2x$$
$$11x = 33$$
$$x = 3$$

Substitute the value for x into the first equation.

$$x + y = 15$$
$$3 + y = 15$$
$$y = 12$$

The numbers are **3 and 12**.

Step 4
Look back. Read the words of the problem again and see whether your answers make sense.

The sum of two numbers is 15. $3 + 12 = 15$ True

Nine times the first number is 3 more than twice the second.

$$9(3) = 3 + 2(12)$$
$$27 = 3 + 24$$
$$27 = 27 \quad \text{True}$$

Writing a system is another tool that you can use to solve problems. Good test takers know that there are many ways to solve a problem. Don't worry about finding the perfect way. Use a method that makes sense to you.

Each of these word problems can be solved by writing a system of equations. Solve using any method. Problems 1 and 2 are started for you.

1. One side of a triangle is 3 times another side. The third side measures 20 inches. The perimeter of the triangle is 52 inches. What are the measures of the sides?

   Let x represent one unknown side.

   Let y represent the other unknown side.

   $$\begin{cases} x = 3y \\ x + y + 20 = 52 \end{cases}$$

2. Lance has $5.95 in coins, all dimes and quarters. If he has 31 coins in all, how many are quarters?

   Let d represent the number of dimes.

   Let q represent the number of quarters.

   $$\begin{cases} d + q = 31 \\ 0.10d + 0.25q = 5.95 \end{cases}$$

3. In a basketball game, the Tigers outscored the Bulldogs by 25 points. If the two teams scored a total of 209 points, how many points did the Bulldogs score?

4. Ben is throwing darts to hit a target at a carnival. If he hits the target, he earns 10 points. If he misses, he loses 15 points. After 40 tries, his score is –50. What percent of his tries hit the target?

5. On the scale, 5 trapezoids have the same weight as 8 pentagons.

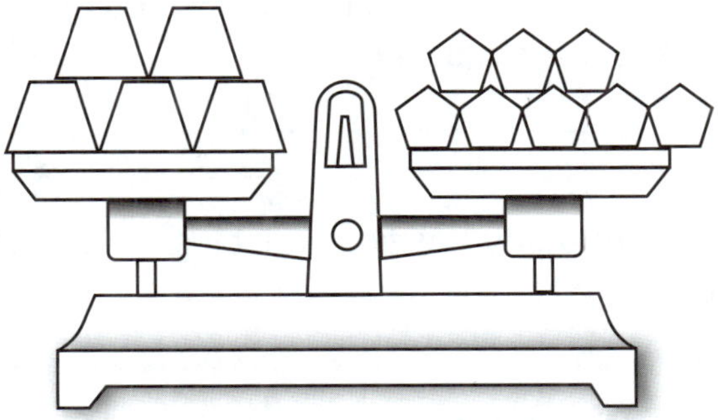

   Together, 2 trapezoids and 1 pentagon weigh 42 grams. How much would 2 pentagons and 1 trapezoid weigh?

6. The sum of the ages of Bob and Cathy is 84 years. In 3 years, Bob will be twice as old as Cathy. How old is Cathy now? (***Hint:*** If b represents Bob's age now, his age in 3 years will be $b + 3$.)

7. A 2-digit number is 4 times the sum of its digits. The ones digit is 3 more than the tens digit. What is the number? (***Hint:*** If t is the tens place and o is the ones place, a 2-digit number can be represented as $10t + o$ and the sum of the digits as $t + o$.)

8. In a fraction, the numerator is 12 less than the denominator. When simplified, the fraction is equal to $\frac{3}{5}$. What is the fraction before it is simplified?

Answers start on page 173.

GRAPH LINEAR INEQUALITIES

An inequality has a range of solutions. When an inequality has only one variable, its solution set can be graphed on a line. In the example below, an open circle shows that the graph starts at –1 but does not include –1. The shaded portion indicates that all the numbers greater than –1 are solutions to the inequality.

$n > -1$

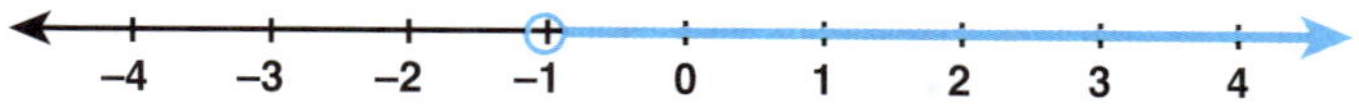

A **linear inequality** has two variables. Study the example below to learn how to graph a linear inequality.

> **TIP**
>
> *See* Math Sense 2: Focus on Problem Solving, *pp. 70–73, for a complete review of graphing and solving inequalities.*

GRAPHING A LINEAR INEQUALITY

Example Graph the solution set to the inequality $y > 2x + 1$.

Step 1

Graph the inequality as though it were a linear equation. Draw the line on the coordinate plane.

- Use a solid line for the symbols ≤ and ≥.
- Use a dashed line for the symbols < and >.

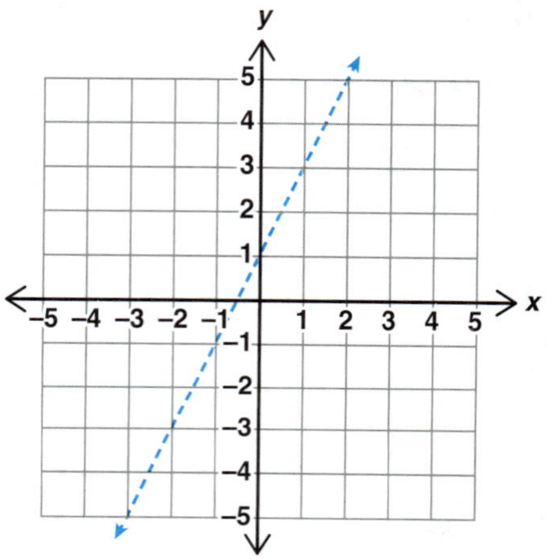

Step 2

The dotted line divides the graph into two regions. One of the regions contains the solutions to the inequality.

Pick a point on one side of the line and test it in the inequality. The origin (0, 0) is often a good choice.

$y > 2x + 1$	Point (0, 0) is not a solution.
$0 > 2(0) + 1$	Shade the region on the opposite
$0 > 1$ False	side of the line.

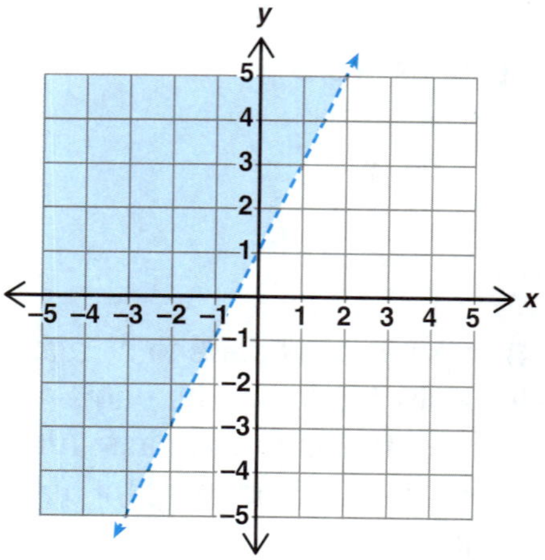

Any point in the shaded region is a solution to the inequality. Points that lie on the line are not a solution because the line is drawn with dashes.

If you are given a graph of an inequality, you can tell whether a point is a solution to that inequality. Remember, if the line is solid, the points on the line are also solutions.

Write a ✓ in a column to show that the point is a member of the solution set for that graph.

	Point	Graph A	Graph B
1.	(0, 0)		
2.	(1, 3)		
3.	(–3, –4)		
4.	(–1, 0)		
5.	(4, –2)		

Graph A

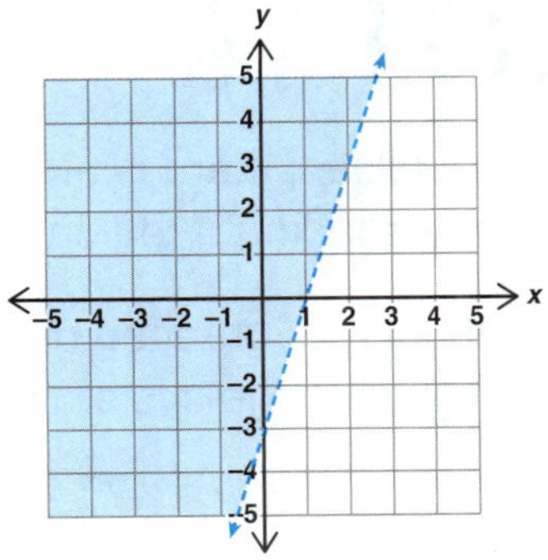

Graph B

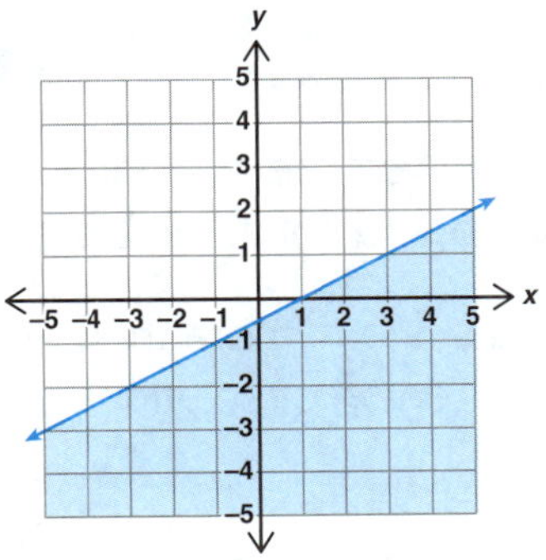

Graph the inequality.

6. $y \geq -2x + 1$

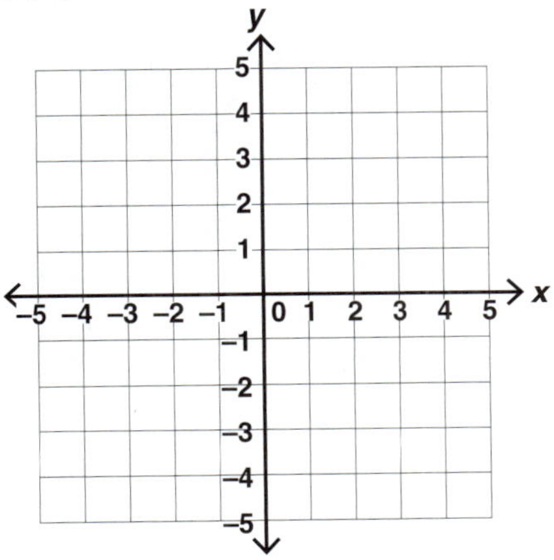

7. $y \geq \frac{2}{3}x - 2$

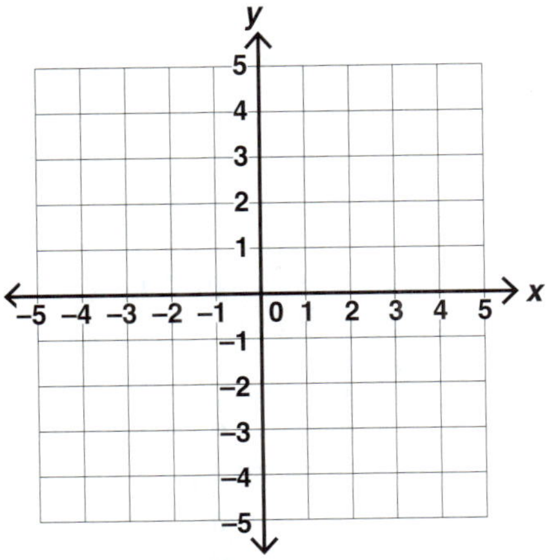

8. $y > x$

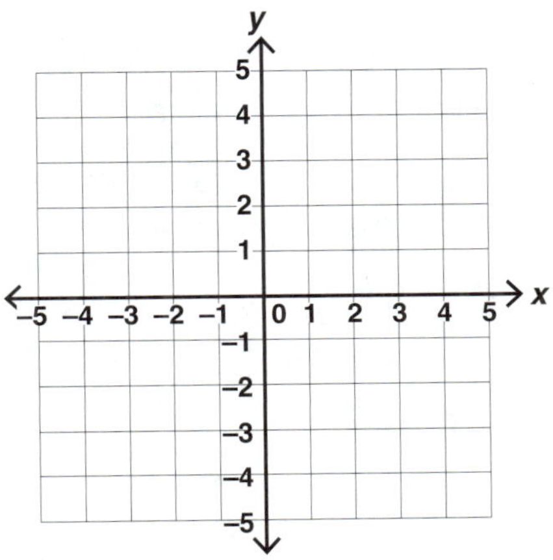

9. **Draw** The solution set of an inequality includes points (–2, 3) and (1, 5), but points (–1, 2) and (4, 4) are not solutions.

On the graph below, draw an inequality that meets these requirements.

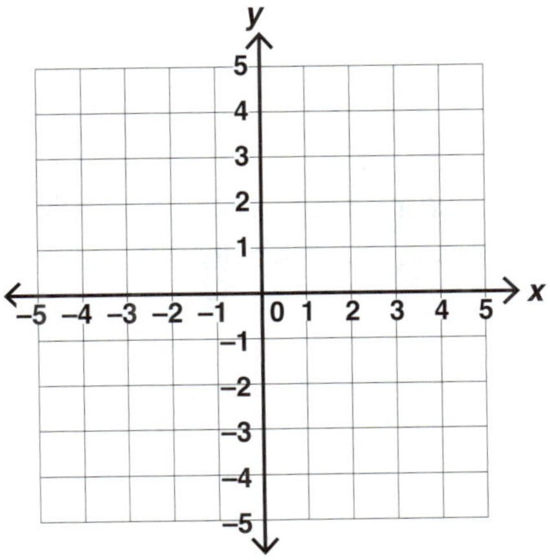

Answers start on page 173.

SOLVE SYSTEMS OF LINEAR INEQUALITIES

When you are working with inequalities, there is always a range of solutions. On a coordinate plane, the solutions are shown by shading a region of the graph.

In a system, more than one inequality is graphed on the same plane. Each inequality has its own region of solutions. The solution to the system is the space where those regions overlap.

GRAPHING A SYSTEM OF LINEAR INEQUALITIES

Example Graph: $\begin{cases} y < -x - 1 \\ y \leq 4x - 3 \end{cases}$

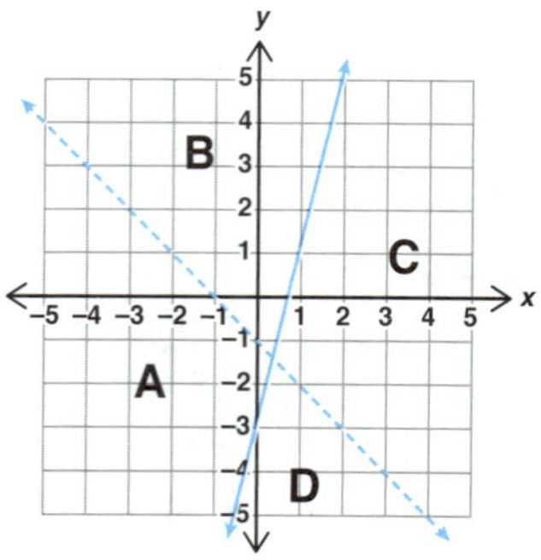

Step 1
Graph both lines on a coordinate plane.

- Use a dashed line if the inequality contains $>$ or $<$.
- Use a solid line if the inequality contains $\geq$ or $\leq$.

The two lines intersect to create four regions. Here they are labeled A, B, C, and D. One of these regions is the solution to the system.

Step 2
Figure out which region is shaded for each inequality. The solution is the space or spaces where the solutions to the individual inequalities overlap.

The graph of $y < -x - 1$ includes regions A and D.

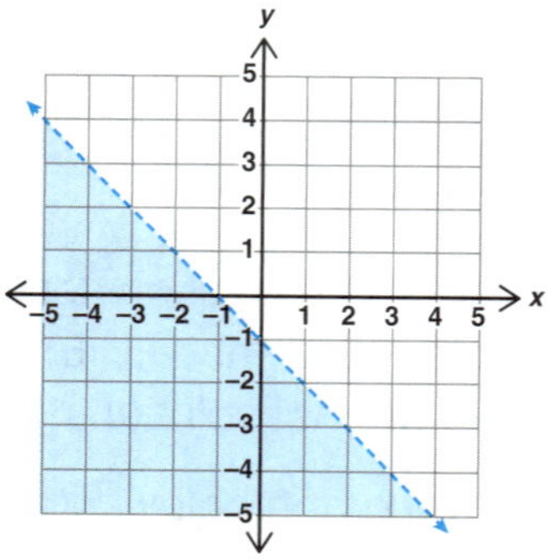

The graph of $y \leq 4x - 3$ includes regions C and D.

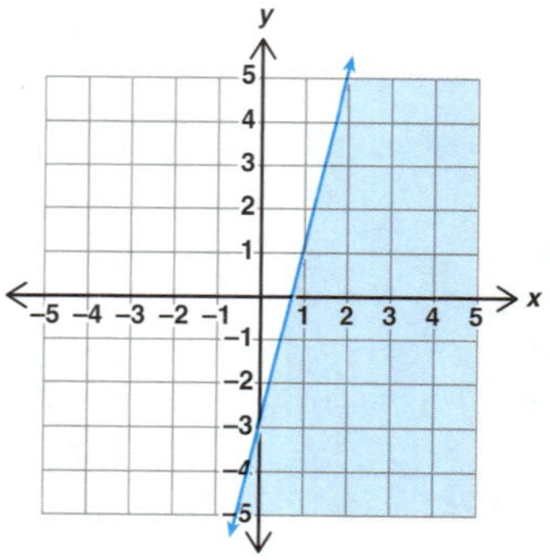

Region D is common to both inequalities. Any point in this region makes both inequalities true.

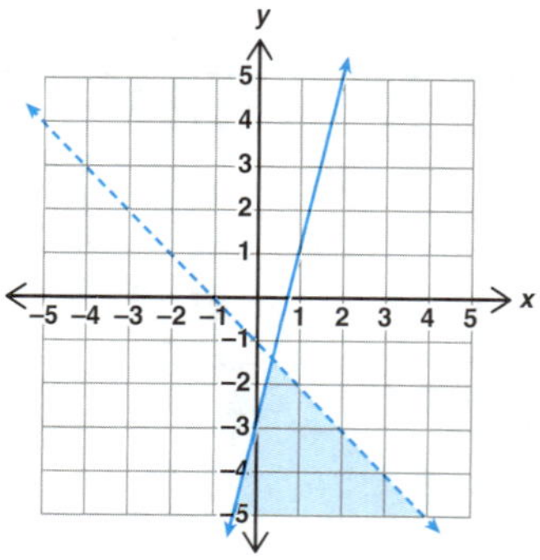

Check: You can check your work by substituting a point from the region that you think is the correct one into the original inequalities. Try (1, –3), a point in the shaded region.

$y < -x - 1$	$y \leq 4x - 3$
$-3 < -(1) - 1$	$-3 \leq 4(1) - 3$
$-3 < -2$ True	$-3 \leq 1$ True

Remember, you can find the correct region by testing points. Choose a point from a region and test it in both inequalities. If the point makes both inequalities true, you know this region is part of the solution. Continue until you have tested each region.

Shade the region that represents the solution set for each system.

1. $\begin{cases} y \geq 3x + 1 \\ y \geq -x - 3 \end{cases}$

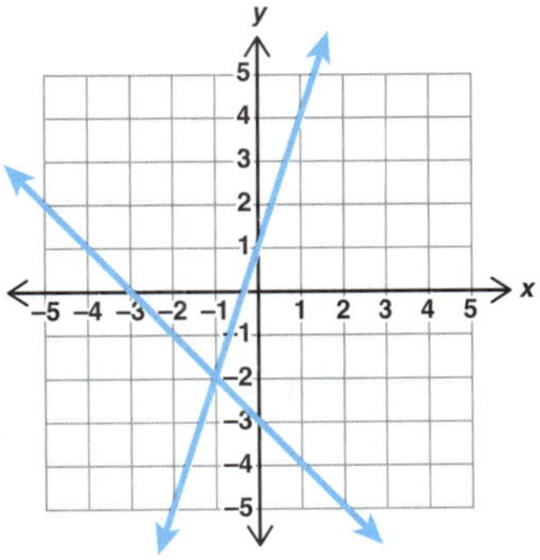

3. $\begin{cases} y < 2x + 1 \\ y \geq -x - 2 \end{cases}$

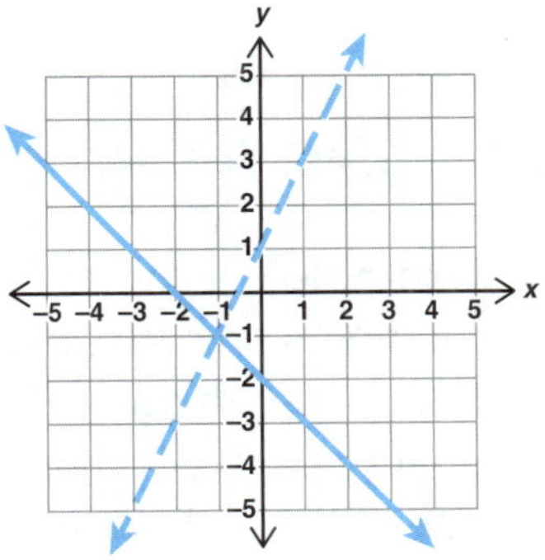

2. $\begin{cases} x \leq -1 \\ y \geq x + 2 \end{cases}$

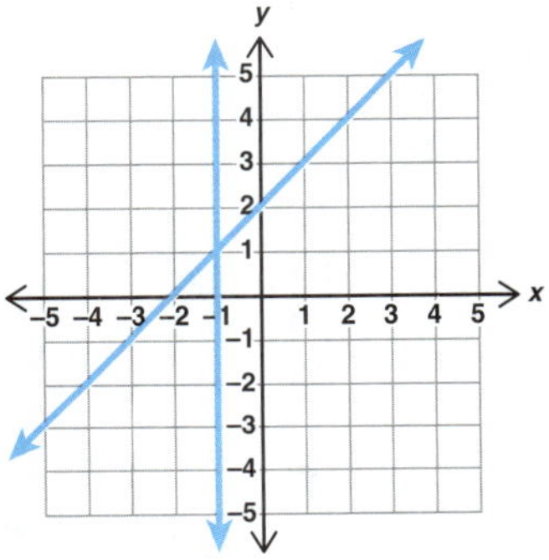

4. $\begin{cases} y < \frac{2}{3}x + 3 \\ y < -x - 2 \end{cases}$

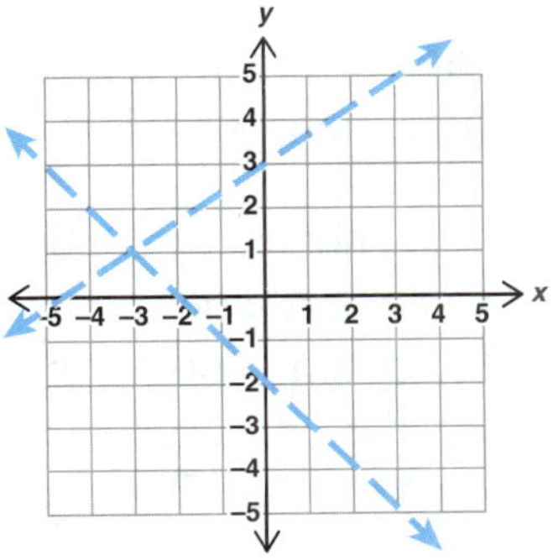

Graph the solution set of each system.

5. $\begin{cases} y < -\frac{1}{2}x + 2 \\ y > x - 1 \end{cases}$

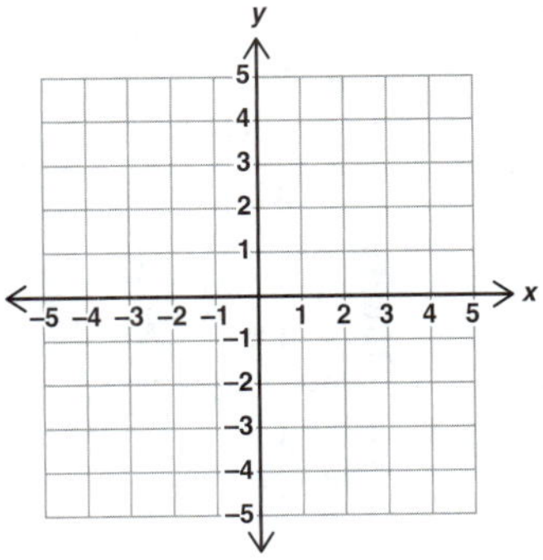

7. $\begin{cases} y < 5x - 3 \\ y \leq -x + 3 \end{cases}$

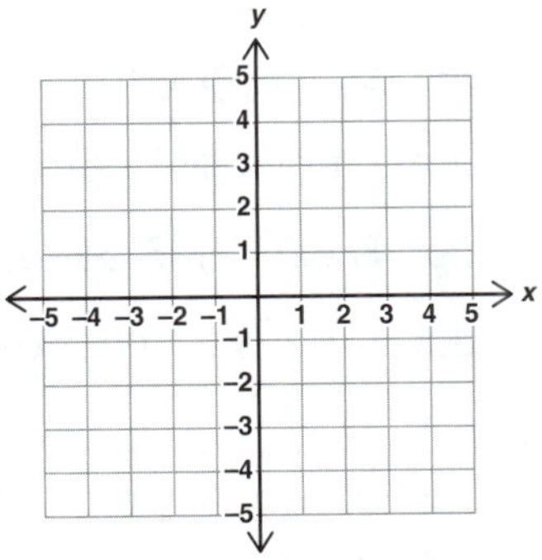

6. $\begin{cases} y \geq x - 2 \\ y > -4x + 3 \end{cases}$

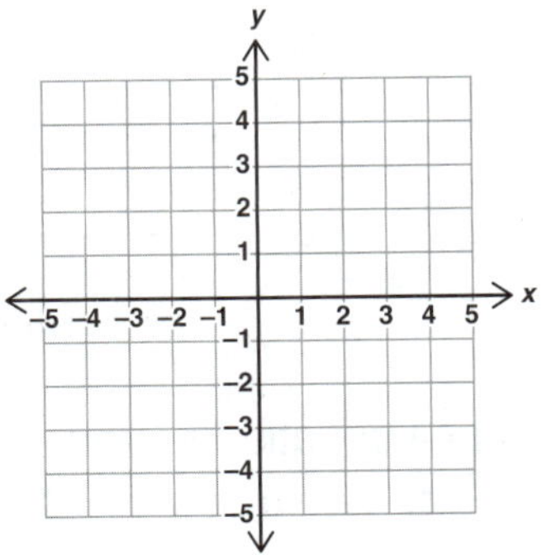

8. $\begin{cases} y < x + 2 \\ y > -3x - 2 \end{cases}$

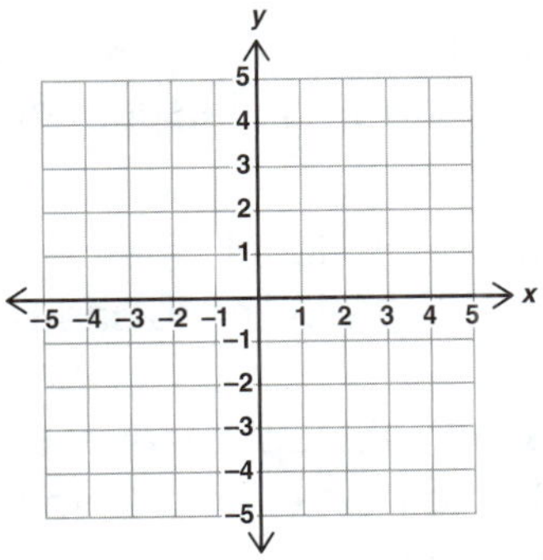

Answers start on page 174.

Problem Solver

INTERPRET GRAPHS OF SYSTEMS

Graphs are often used to help businesses and individuals make important decisions. They can be used to find the best economic strategies and to optimize profits.

USING A GRAPH TO MAKE DECISIONS

Example A swim club offers two membership options. Plan A charges \$8 per month and \$5 per visit. Plan B charges \$40 per month for unlimited use of the club. The graph shows both plans.

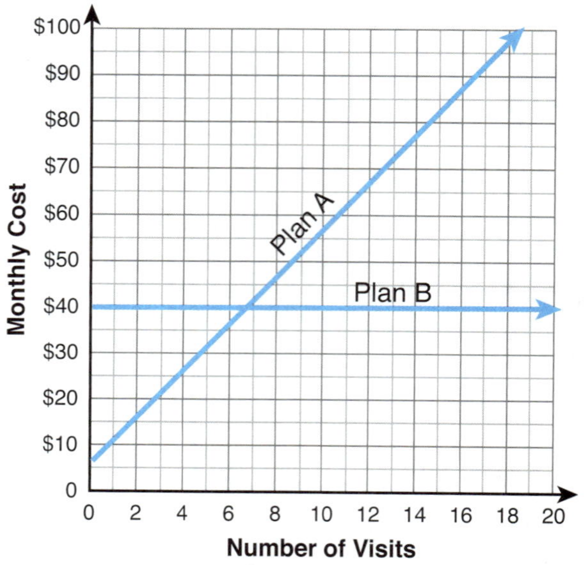

Plan B becomes the better deal after how many visits?

The lines intersect between 6 and 7 visits. If a member uses the club 6 or fewer times per month, Plan A is best. If a member visits **7 or more times per month**, Plan B is best.

Problems 1–3 refer to the graph and information below.

John is looking for a sales job. Three job postings are shown on the graph.

- Job A pays \$60 per day.
- Job B pays \$45 per day plus \$2.50 per sale.
- Job C pays \$10 per day plus \$5 per sale.

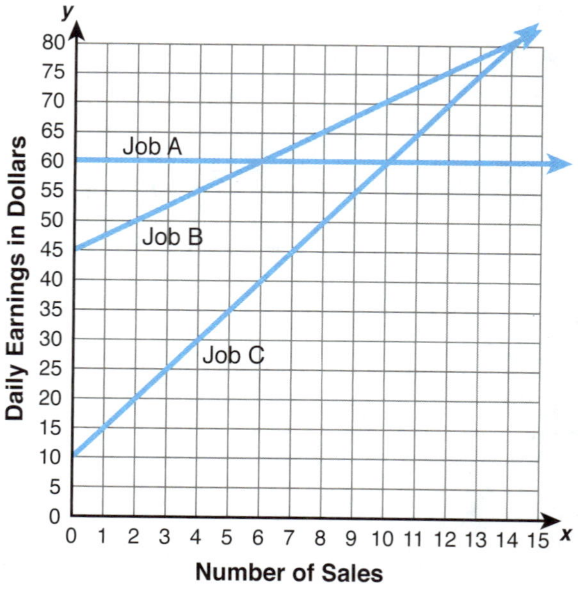

1. Which job would pay the most if John averaged 5 sales per day?

2. How many sales would John need per day to make Job C the best opportunity?

3. John doubts that he can make more than 12 sales per day. Which job should he pursue? Why?

When you optimize, you are attempting to use all the factors in a situation to your advantage. In the next example, you will see how the graph of inequalities can help you make an optimal decision.

USING A GRAPH TO OPTIMIZE

Example Ethan needs waterproof boxes to store his belongings. His storage unit has a maximum of 72 square feet of floor space. He can afford to spend up to $215. The boxes should give him the maximum volume, or cubic feet of storage.

The hardware store has two box sizes.

- Box A is $15, covers 6 square feet of floor space, and holds 12 cubic feet.
- Box B is $30, covers 8 square feet of floor space, and holds 20 cubic feet.

How many should Ethan buy of each box? How many cubic feet of storage space will he have?

The graph is a system of inequalities that shows all the possible solutions.

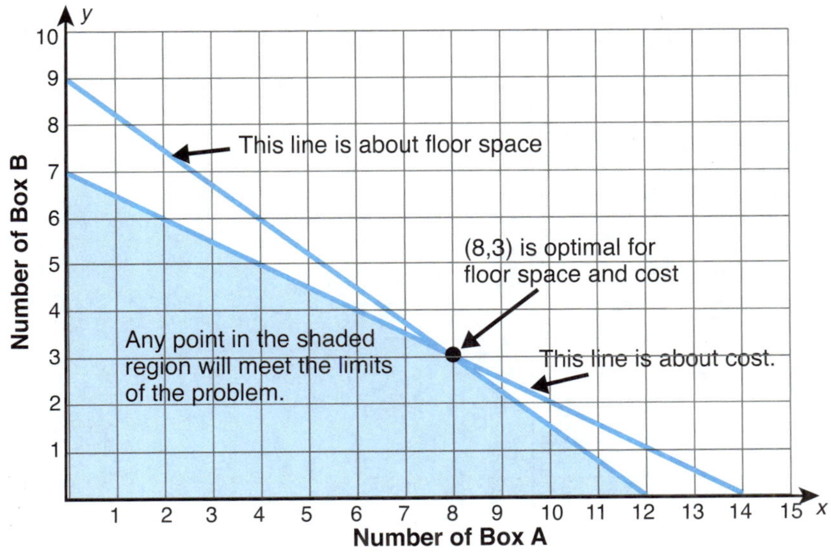

You know that the total volume equals $12x + 20y$.

$$12x + 20y = 12(8) + 20(3)$$
$$= 96 + 60$$
$$= 156 \text{ cubic feet}$$

The optimal choice is where the lines intersect (8, 3). Ethan should buy **8 of Box A and 3 of Box B**, which will give him 156 cubic feet of storage.

Use the information and graph to answer problems 4–6.

a. A school store expects to sell 40 to 60 T-shirts and 25 to 40 sweatshirts per week.

b. The school can print no more than 90 shirts in all per week.

c. The school makes an $8 profit on each sweatshirt and a $15 profit on each T-shirt.

Let *x* equal the number of T-shirts and *y* equal the number of sweatshirts.

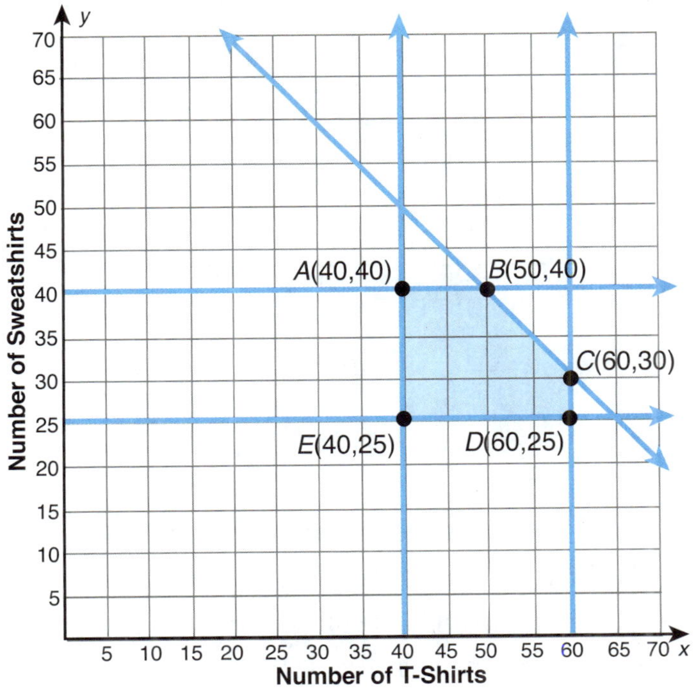

4. The number of sweatshirts, *y*, is expressed by the inequality $25 \leq y \leq 40$. Write an inequality for the number of T-shirts.

5. Write an inequality to represent Fact b.

6. The equation $P = \$15x + \$8y$ is used to find total profit (P). Test each corner of the shaded region in the equation. What is the optimal number of T-shirts and sweatshirts the school should print per week to obtain the greatest profit?

Answers start on page 175.

Unit 3 Review

Solve as directed.

1. A system of equations is shown on the graph.

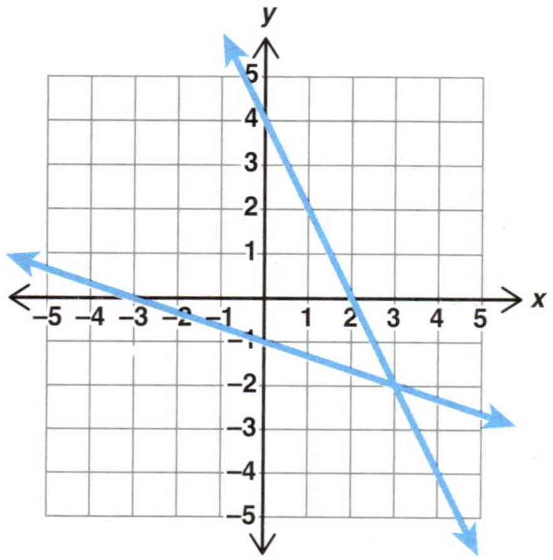

Write the solution to the system in the boxes below.

$x =$ ☐ $y =$ ☐

2. The graph shows the following system:

$$\begin{cases} y = \frac{1}{2}x - 2 \\ y = \frac{1}{2}x + 4 \end{cases}$$

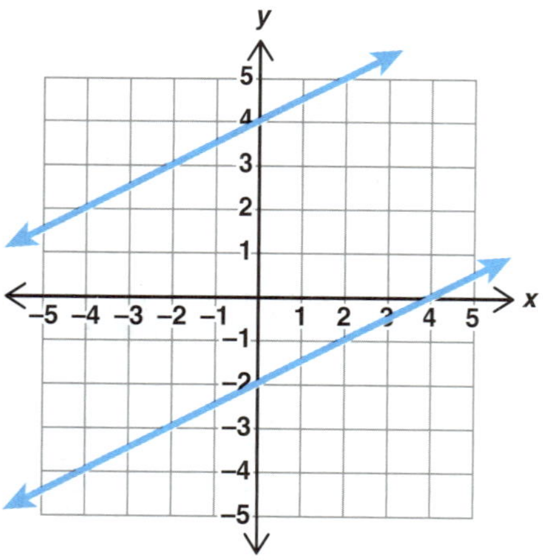

What can you conclude about the system?

A. This system cannot be graphed.

B. The solution is not shown on the graph.

C. There is no solution.

D. There are an infinite number of solutions.

3. Which ordered pair is the solution to the system?

$$\begin{cases} -2x + y = -7 \\ -4x - 3y = 11 \end{cases}$$

A. (–2, –1)

B. (3, –1)

C. (2, –3)

D. (1, –5)

4. Write the solution to the system in the box.

$$\begin{cases} 4x - 3y = 6 \\ -7x + 8y = -16 \end{cases}$$

☐

5. Dylan notices something odd about the following system:

$$\begin{cases} -5x + y = 6 \\ 10x - 2y = -12 \end{cases}$$

He notices that if you multiply every term in the first equation by –2, the first equation becomes exactly the same as the second equation.

What can you conclude about this system? Explain your reasoning.

☐

6. Graph the system of inequalities on the coordinate plane to the right.

$$\begin{cases} y \geq -x - 1 \\ y \geq x - 3 \end{cases}$$

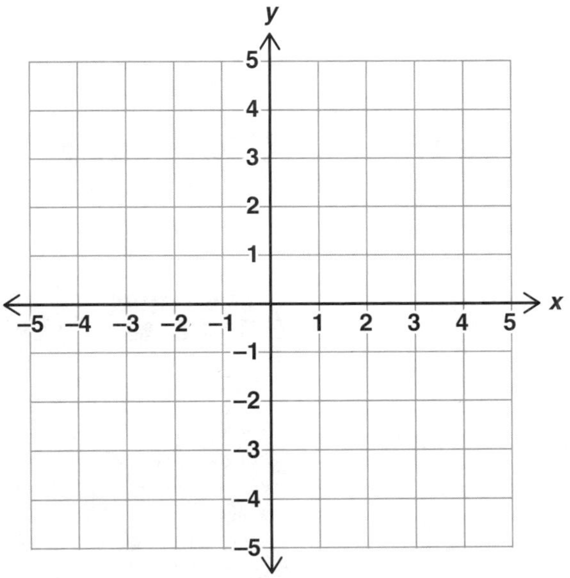

7. Is the origin (0, 0) a member of the solution set? Write Yes or No.

8. A hospital orders 32 new shelving units to store patient files. Some of the units have 5 shelves and some have 4 shelves. The new units will provide a total of 145 shelves to file storage. How many 5-shelf units did the hospital buy?

☐ 5-shelf units

9. Extreme Bowling charges \$5 per game and \$4 to rent a pair of shoes. Lakeside Bowling charges \$4 per game and \$8 to rent a pair of shoes. How many games would you need to bowl in order for both bowling alleys to charge the same amount?

A. 2

B. 3

C. 4

D. 5

10. Which of the ordered pairs is in the solution set of the system below?

$$\begin{cases} y \leq -6x + 3 \\ y > -x - 2 \end{cases}$$

A. (–3, 1)

B. (–2, 3)

C. (0, –4)

D. (2, 4)

11. The tenants of an apartment building are planning to make a rectangular community garden. They have enough money to buy no more than 120 feet of fencing. They would like the length (*L*) of the garden to be less than twice the width (*W*).

In the boxes below, write a system of inequalities to represent the information in the problem.

12. On the first day of a sale at Runway Clothing, all jeans were \$25 and all shirts were \$22. The store took in \$686.

On the second day, the store reduced the price of both jeans and shirts to \$20 each. It sold exactly the same number of each item as it had on the first day.

If the store made \$580 on the second day, how many shirts did they sell *in all* during the two-day sale?

A. 13

B. 16

C. 26

D. 32

Answers start on page 175.

UNIT 4

FUNCTIONS

SKILLS

- Patterns and sequences
- Finding the *n*th term
- Working with a geometric series
- Understanding functions
- Reading graphs of functions
- Finding the domain and range of a function

TOOLS AND STRATEGIES

- Comparing functions
- Working with functions in the real world

EVERY PATTERN you see is formed using a rule. A rule can be written in words. For example, a rule to make a wallpaper pattern could read: "Alternate blue and white vertical stripes. The blue stripes are two inches wide, and the white stripes are one inch wide."

A **function** is a rule written using algebra. A function tells how one quantity depends on other quantities. One way to write a function is as a formula. For example, the area of a rectangle depends on its length and width: $A = lw$. The number of miles you drive depends on how fast you drive and how many hours you drive: $d = rt$.

In this unit, you will learn how to use functions to describe mathematical patterns and relationships. You will learn how to analyze graphs of functions and how to work with different kinds of patterns and sequences.

USING FUNCTIONS

You use functions when you:

- Find area and perimeter of geometric figures
- Calculate your gas mileage
- Figure out how much a number of items will cost
- Find a missing number in a pattern
- Analyze data displayed in graphs and charts

Read the following items. Describe some of your experiences in writing or discuss them with a partner.

1. Frames, borders, and tiles often have patterns. Find a geometric pattern in your home or community. Write a rule that describes the pattern. Share what you wrote with a friend. Does your friend have enough information to continue the pattern?

2. Find the formula for simple interest on the formula sheet in the Toolkit, p. 192. What factors does simple interest depend on?

3. Many common machines are programmed with functions. What are some of the services, or functions, available at an ATM machine?

4. Patterns can also involve numbers. A palindrome is a number that is the same whether it is read forward or backward, such as 343 or 10101. How many 3-digit palindromes are there? How would you organize your work so that you could count them? Do you see any patterns in your organization?

MATH TALK

How can you tell how old a dog is in people years? Most people say that you multiply the dog's age by 7.

However, experts describe the process this way. If your dog is 1 or 2 years old, multiply the age by 10.5. If your dog is older than 2 years, multiply the first 2 years by 10.5 and the remaining years by 4. Then add 13 years.

Make a table that connects dog years to people years for the first 10 years of a dog's life. Try both methods. How do the methods compare? Both methods are functions. Why do you think that people prefer the first method of multiplying by 7?

PATTERNS AND SEQUENCES

Mathematics has been defined as the science of patterns. Everything from art to nature to construction projects involves patterns.

A **pattern** has one or more repeating elements. The numbers in the list below are the multiples of 8. Focus on the ones place in each number. What do you notice?

8, 16, 24, 32, 40, 48, 56, 64, 72, 80, 88, 96, 104, 112, 120, . . .

The digits in the ones place form a pattern that repeats. Every 5 numbers, the pattern starts again: 8, 6, 4, 2, 0—8, 6, 4, 2, 0—8, 6, 4, 2, 0, and so on.

A **sequence** is a pattern that is defined by a rule. Look at the following sequence. What do you notice? How would you describe the sequence?

1, 6, 11, 16, 21, 26, 31, 36, 41, 46, 51, . . .

You may have noticed that the ones place forms a pattern: 1, 6, 1, 6, 1, 6, . . . The sequence is formed by adding 5. You could say, "The sequence starts at 1 and counts by 5s." This statement describes the rule that forms the sequence.

A sequence is **finite** if it has a limited number of terms. A sequence is **infinite** if the terms go on forever. Each number or other item in the sequence is called a **term**.

To talk about a sequence, you need to be able to identify a term's **position** in the sequence and its **value**. In the sequence 5, 10, 15, 20, 25, . . . 50, the 3rd term is 15. The position is 3rd; the value is 15.

FINDING WHAT COMES NEXT

Example What number comes next in the sequence?

Position	1st	2nd	3rd	4th	5th	6th
Value	1	2	4	8	16	?

Examine the sequence, which is shown in the row of values. What is happening from one term to the next? The 2nd term is 2 times the 1st term. The 3rd term is 2 times the 2nd term. Each term is 2 times the term that came before.

The 6th term must be 2 times the 5th term: $16 \times 2 = 32$. The next term is **32**.

Patterns and sequences can contain letters, symbols, or geometric figures. To solve a problem with a pattern, look for something that repeats and continue the pattern. To solve a problem with a sequence, think about what changes from one term to the next.

CONTINUING A PATTERN WITH SHAPES OR SYMBOLS

Example Which choice is the next term in the pattern?

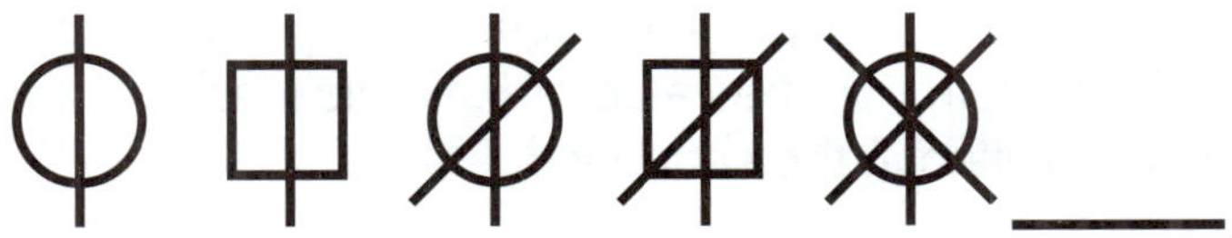

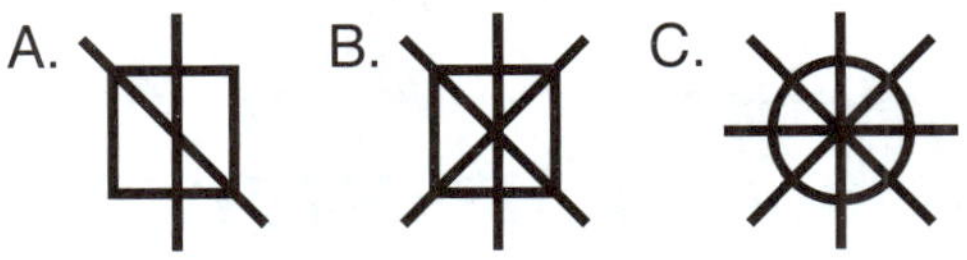

The pattern is alternating circles and squares, so the next term needs to be a square. The first pair of terms has one line drawn through the circle or square, the second pair has two lines, and the last circle has three lines. The next square needs three lines.

Choice B is correct because it is a square with three lines.

Draw or describe the next item in each sequence.

1. 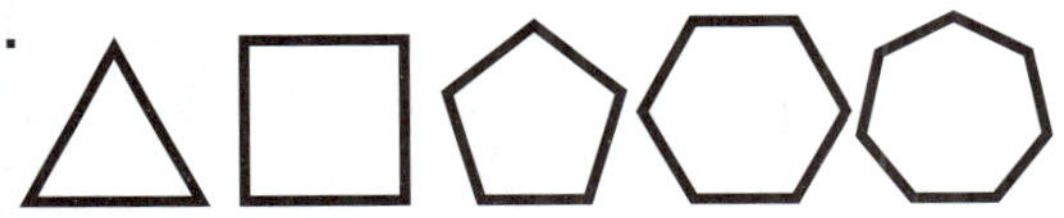 ______

2. A B C D E F G H I ______

3. A, E, I, M, Q, ______

4. 100 91 82 73 64 ______

Solve as directed.

5. **Explain** If the pattern to the right continues, how many blocks would you expect to see in the next figure? Explain your thinking.

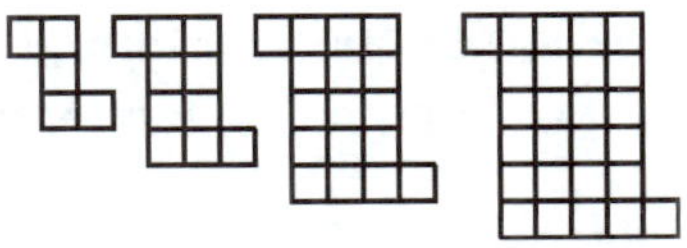

6. If the pattern continues, what symbol would be the 50th term?

 ... ______

7. If the fraction $\frac{7}{11}$ is changed to a decimal, what digit will be in the 100th decimal place? (***Hint:*** Change the fraction to a decimal using your calculator.)

8. What time is missing from the sequence?

Answers start on page 176.

FINDING THE *n*TH TERM IN A SEQUENCE

Each term in a sequence is identified by its place in the sequence: 1st, 2nd, 3rd, etc. The *n*th term can represent any position, depending on the value of *n*. In this lesson, you will see how to find a specific term based on its position in the sequence.

In an **arithmetic sequence**, the terms always go up or down *by the same amount.*

APPLYING A RULE TO MAKE AN ARITHMETIC SEQUENCE

Example 1 Create a sequence of six numbers using this rule. $8n - 13$

Step 1
Make a table with the numbers from 1 to 6. These become the values for *n*.

n	1	2	3	4	5	6
$8n - 13$	–5	3	11	19	27	35

Step 2
Apply the rule $8n - 13$ to each number. What do you notice about the sequence? The terms go up by 8.

The sequence is **–5, 3, 11, 19, 27, 35**.

If you are given a sequence, you can work backward to find the rule.

WRITING A FUNCTION RULE FOR A SEQUENCE

Example 2 Find the rule that forms this sequence. 9, 16, 23, 30, 37, . . .

Step 1
Examine the sequence. Find the difference between the terms. Is it always the same? Yes, this pattern goes up by 7.

9 16 23 30 37
7 7 7 7

The difference of 7 tells you that the sequence is based on the multiples of 7. The variable *n* represents a term's position in the sequence. Multiply every value of *n* by 7.

n	1	2	3	4	5
$7n$	7	14	21	28	35

Step 2
The resulting sequence (7, 14, 21, . . .) doesn't equal the original sequence. You need to adjust the sequence by adding 2 to each term.

+2	9	16	23	30	37

These steps transform the numbers for *n* into the original pattern. Therefore, the rule that forms this sequence is $\mathbf{7n + 2}$.

If the pattern in the example had gone down by 7, you would have multiplied by –7.

Once you know the rule, you can find the nth term.

FINDING THE nTH TERM

Example What is the 40th term in this sequence? 11, 14, 17, 20, 24, . . . _____

Step 1
Find the rule.

The difference between terms is 3.

Add 8 to adjust. The rule is $3n + 8$.

n	1	2	3	4	5
$3n$	3	6	9	12	15
+8	11	14	17	20	24

Step 2
Find the 40th term. Substitute 40 for n.

$3n + 8$
$= 3(40) + 8$
$= 128$

The 40th term is **128**.

Write a sequence of five terms for each rule.

1. $5n - 9$

2. $6(n - 11)$

3. $9n + 4$

4. $n - (2n + 5)$

Write the rule for each pattern.

5. –2, –5, –8, –11, –14, . . .
(***Hint:*** The sequence goes down by –3, so multiply n by –3.)

6. 13, 15, 17, 19 21, . . .

7. –3, –13, –23, –33, –43, . . .

8. –7, 5, 17, 29, 41, . . .

Find the nth term as directed.

9. Find the 10th term in the sequence.
–6, –10, –14, –18, . . . ________

10. Find the 25th term in the sequence.
15, 28, 41, 54, . . . ________

11. Find the 40th term in the sequence.
–8, –1, 6, 13, . . . ________

12. Find the 100th term in the sequence.
100, 75, 50, 25, . . . ________

13. Explain A sequence begins: 8, 14, 20, ... Doug is asked to find the 30th term in the sequence. He reasons, "The 3rd term is 20, so the 30th term must be 10 times 20, or 200." What is wrong with Doug's thinking?

Answers start on page 176.

WORKING WITH A GEOMETRIC SEQUENCE

In the last lesson, you explored arithmetic sequences. In an arithmetic sequence, the difference between consecutive terms is always the same. The next term in the sequence is found by adding (or subtracting) the difference between terms.

Examples of Arithmetic Sequences

11, 21, 31, 41, . . .	Increases by 10.	The rule is $10n + 1$.
30, 26, 22, 18, . . .	Decreases by 4.	The rule is $-4n + 34$.

In a **geometric sequence**, the ratio between any two consecutive terms is the same. The ratio is found by dividing one term by the other. The next term in the sequence is found by multiplying by that ratio.

FINDING THE NEXT TERM IN A GEOMETRIC SEQUENCE

Example What term comes next in this sequence? 2, –4, 8, –16, 32, . . .

Step 1

Figure out what kind of sequence it is. Find the difference of the terms. The sequence is not arithmetic because the difference isn't always the same.

2 –4 8 –16 32

–6 +12 –24 +48

Step 2

Find the ratio of consecutive terms. Write a ratio of the second term to the first term and simplify. Continue with other terms. Always write the ratios in the same order.

$\frac{-4}{2} = -2$ $\frac{8}{-4} = -2$

$\frac{-16}{8} = -2$ $\frac{32}{-16} = -2$

The next term in the sequence is always found by multiplying the preceding term by –2.

The sequence is geometric.

Step 3

Use the ratio to find the 6th term.

$32 \cdot -2 = -64$

The next term is **–64**.

Find the next term for each geometric sequence.

1. 5, 10, 20, 40, 80, ________

2. 512, 256, 128, 64, 32, ________

3. 1, –3, 9, –27, 81, ________

4. 100, 10, 1, 0.1, 0.01, ________

To find the *n*th term in a geometric series, multiply the first term by the ratio raised to a power that is 1 less than *n*.

FINDING THE *n*TH TERM IN A GEOMETRIC SEQUENCE

Example What is the 7th term in this geometric sequence? 3, 12, 48, 192, . . . ______

Step 1
Find the ratio of consecutive terms. $\frac{12}{3} = 4$ $\frac{48}{12} = 4$

The next term in the sequence is always the preceding term multiplied by 4.

Step 2
By the 7th term, you will have multiplied 6 times. The number of times is always 1 less than the number of terms. Multiply the *first* term, 3, by 4^6.

Use your calculator:
3 × 4 ^ 6 [enter] $3*4^6$ 12288

The 7th term is **12,288**.

Decide whether each sequence is arithmetic or geometric.

5. 8, 18, 28, 38, . . . arithmetic / geometric

6. $\frac{1}{2}, \frac{1}{4}, \frac{1}{8}, \frac{1}{16}, \ldots$ arithmetic / geometric

7. 6, –6, 6, –6, . . . arithmetic / geometric

8. –4, –10, –16, –22, . . . arithmetic / geometric

Find the next term in the sequence.

9. –4, –8, –16, –32, ______

10. 1,000, 200, 40, ______

Find the *n*th term as directed.

11. Find the 7th term in the sequence.
2, 6, 18, 54, . . . ______

12. Find the 8th term in the sequence.
$\frac{1}{10}$, 1, 10, . . . ______

13. Find the 9th term in the sequence.
0.15, 0.3, 0.6, 1.2, . . . ______

14. Find the 7th term in the sequence.
1, 4, 16, 64, . . . ______

Answers start on page 177.

FUNCTION BASICS

A set of ordered pairs is a **relation**.

The first member in each ordered pair is part of the **domain**. In this set, 2, 3, and 4 are in the domain. The second member is part of the **range**. The range consists of 4, 6, and 8.

{(2, 4), (3, 6), (4, 8)} ← A set is written in braces.

A relation is a **function** when each element of the domain is paired with *exactly one* element of the range.

The relation on the far right is *not* a function because 6 is paired with two different elements in the range.

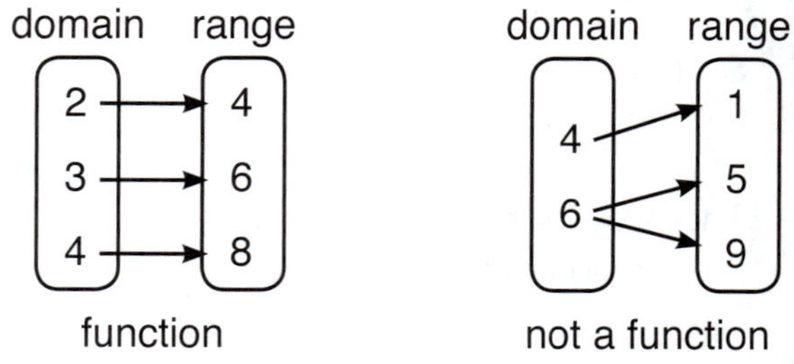

A function is like a machine that applies a rule to each member of the domain. A member of the domain enters the machine. The rule is applied, and a member of the range exits the machine. Here, 3 becomes 6 when the rule "multiply by 2" is applied.

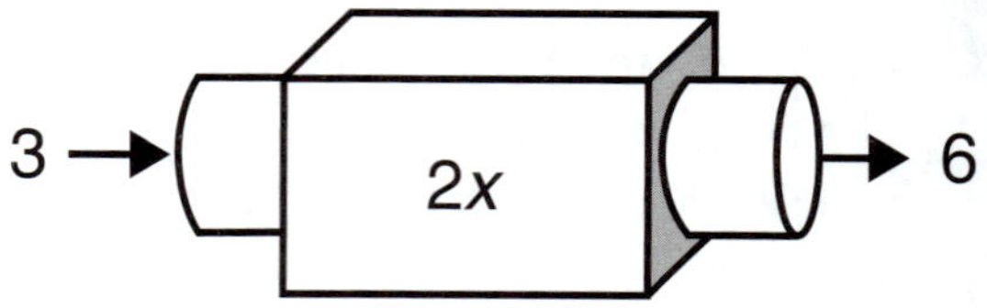

Functions use a special system of notation. The expression $f(x)$ is read "f of x." It means "the function of x." A function states the rule to apply to any value of x. ***Note:*** The letters f, g, and h are usually reserved for the naming of functions.

An example of a function is $f(x) = x^2 - 5$. It says that for every value of x, the function of x is equal to x squared minus 5.

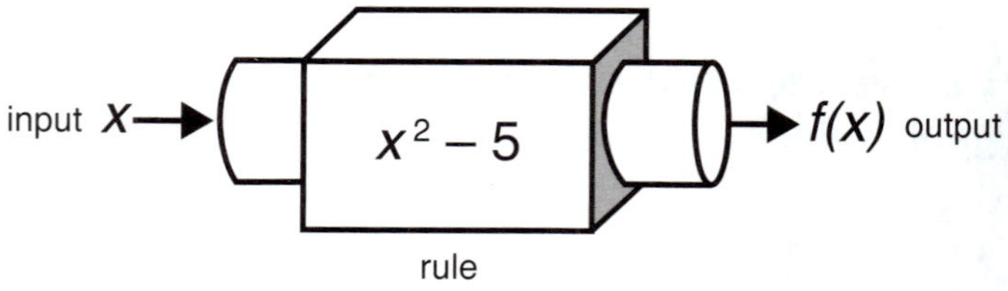

EVALUATING A FUNCTION FOR A GIVEN VALUE

Example For $f(x) = x^2 - 5$, find $f(-2)$ and $f(4)$.

Find $f(-2)$.

Substitute –2 for x in the function and solve.

$$\begin{aligned} f(-2) &= (-2)^2 - 5 \\ &= 4 - 5 \\ &= -1 \end{aligned}$$

Find $f(4)$.

Substitute 4 for x in the function and solve.

$$\begin{aligned} f(4) &= 4^2 - 5 \\ &= 16 - 5 \\ &= 11 \end{aligned}$$

$\boldsymbol{f(-2) = -1}$ and $\boldsymbol{f(4) = 11}$

In some situations, you may need to substitute an expression for x. When you are asked to evaluate a function, substitute whatever is in the parentheses for x. Then simplify the expression.

USING AN EXPRESSION IN A FUNCTION

Example For $g(x) = 3x + 5$, find $g(4n)$.

Substitute $4n$ for x in the function and simplify.

$$g(4n) = 3(4n) + 5$$
$$= 12n + 5$$

$\mathbf{g(4n) = 12n + 5}$

Evaluate each function for the given value and simplify.

1. $f(x) = -4x + 3$

a. Find $f(4)$.

b. Find $f(-10)$.

2. $g(n) = n^3 + 1$

a. Find $g(-2)$.

b. Find $g(2)$.

3. $f(n) = 2n^2 + 5n$

a. Find $f(-3)$.

b. Find $f(9)$.

4. $g(x) = -x + 13$

a. Find $g(-10)$.

b. Find $g(10)$.

5. $h(x) = |\ 2x - 1\ |$

a. Find $h(-5)$.

b. Find $h(5)$.

(***Hint:*** The bars indicate absolute value. Absolute value makes the value inside the bars positive. For a review of absolute value, see *Focus on Problem Solving,* Unit 1, of this series.)

6. $h(n) = |\ n\ | - 6$

a. Find $h(-7)$.

b. Find $h(7)$.

7. For $f(x) = -2x + 1$, find $f(n - 2)$. Remember to simplify the expression.

8. For $g(x) = 3x - 2$, find $g(-2a)$.

9. For $h(n) = 4n - 3$, find $h\left(\frac{b}{4}\right)$.

10. Challenge Start with the function $f(n) = 4n - 5$. If $f(x - 1) = 7$, what must x equal? (***Hint:*** Start by substituting $x - 1$ for n.)

Answers start on page 177.

READING GRAPHS OF FUNCTIONS

Like an equation, a function can be graphed on a coordinate grid.

EVALUATING A FUNCTION USING A GRAPH

Example For $f(x) = 2x - 3$, find $f(2)$.

Read the graph to find $f(2)$. What is the y-coordinate of the line when the x-coordinate is 2?

The y-coordinate is 1, so $\boldsymbol{f(2) = 1}$.

The answer makes sense.

Try substituting 2 for x.

$$\begin{aligned} f(2) &= 2(2) - 3 \\ &= 4 - 3 \\ &= 1 \end{aligned}$$

The answer is the same.

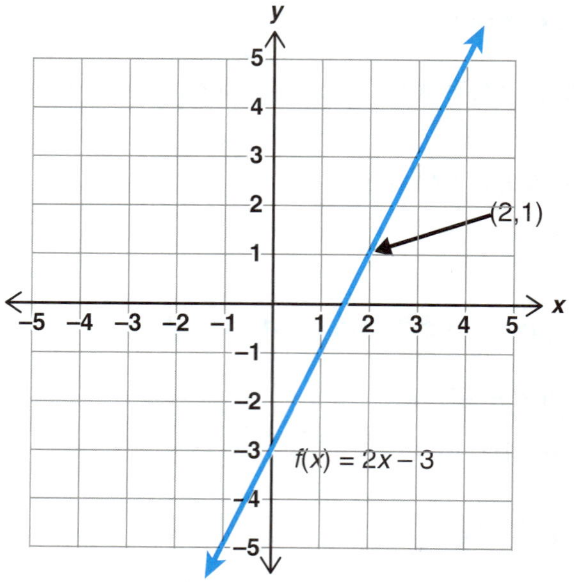

A graph can also help you decide whether a relation is a function. Remember, if a relation is a function, every input can have only one output. That means that a graph of a function will have only one y-coordinate for every x-coordinate.

The **vertical line test** is a quick way to judge whether a graph represents a function. If you can draw a vertical line that passes through more than one point anywhere on a graph, the graph is not a function.

USING THE VERTICAL LINE TEST

Example On each graph, the blue line represents a relation. Is the relation a function?

Imagine drawing a vertical line through each graph.

If the vertical line passes through the graph more than once, then there is an x-coordinate with more than one possible y-coordinate.

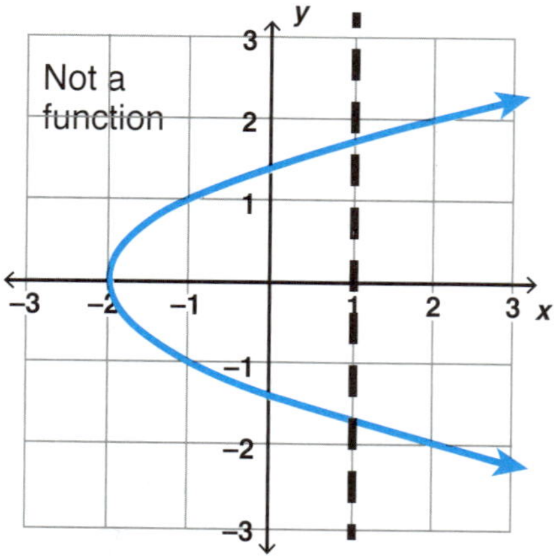

The vertical line crosses the blue line more than once.

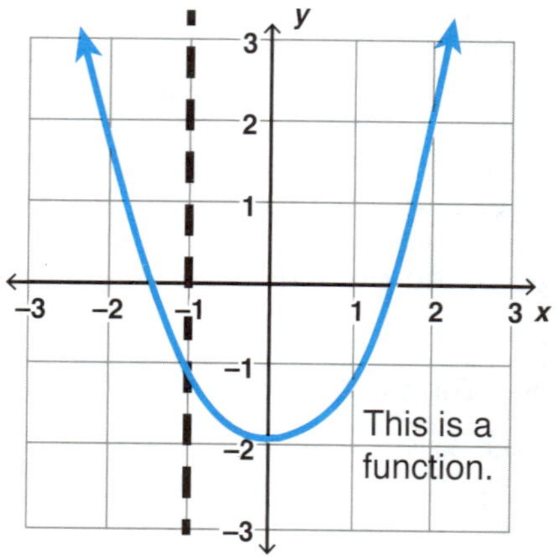

No matter where you place the vertical line, it will cross the blue line only once.

You may be asked a question about a specific interval on a graph. An **interval** is the distance between two points, usually on the *x*-axis.

The graph of function $h(x)$ has several parts. Notice how intervals can be used to describe and analyze the graph.

- During the interval $x \le -1$, $h(x)$ increases as x increases.
- In the interval $-1 \le x < 3$, the function reaches its highest point when $x = 1$.
- The function does not change in the interval $x \ge 3$.

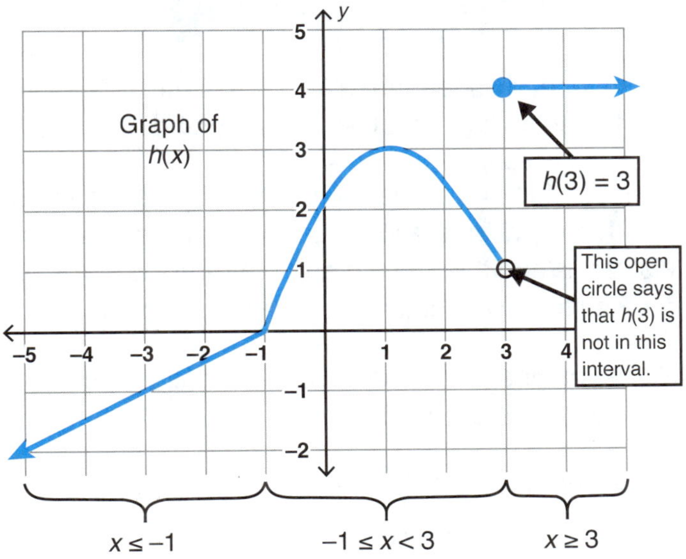

Use the vertical line test.

1. Which graph is not a function?

A.

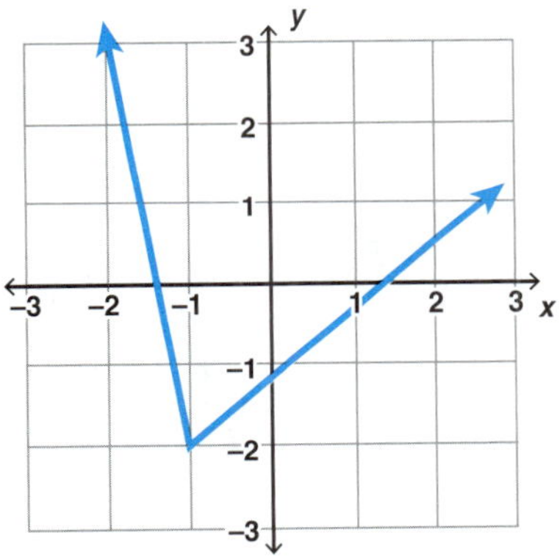

B.

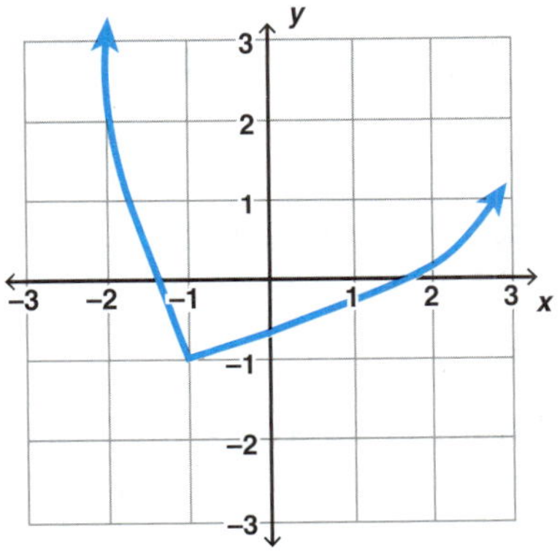

C.

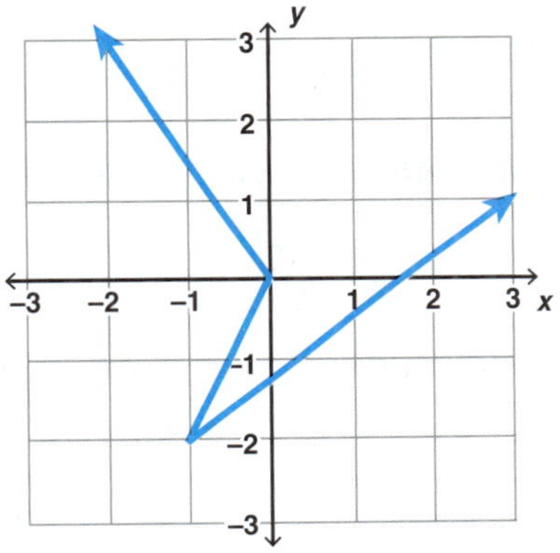

Problems 2–6 refer to the graph on the right.

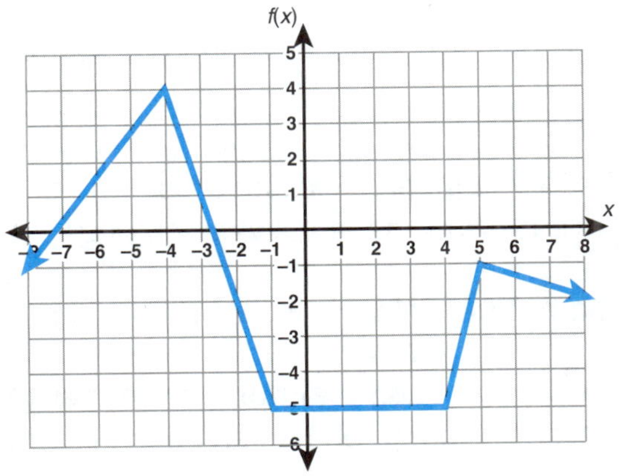

2. What is the value of $f(-2)$?

3. What is the value of $f(5)$?

4. What is the value of x when $f(x)$ is equal to 4?

5. Which interval shows the greatest amount of change?

A. $-8 \le x \le -4$

B. $-4 \le x \le -1$

C. $4 \le x \le 5$

D. $5 \le x \le 8$

6. **Explain** Katie makes the claim that the graph shows the greatest *rate* of change when $4 \le x \le 5$. The rate of change is different from the amount of change. Think about what is meant by a rate. Do you agree or disagree with Katie's claim? Explain your reasoning.

Answers start on page 177.

FINDING DOMAIN AND RANGE OF A FUNCTION

You've already seen that a function is much like a machine. A number from the domain is substituted for the variable, and a number from the range comes out.

For many functions, the domain and the range will be the **set of real numbers**, which includes all rational and irrational numbers. Mathematicians use the symbol $\mathbb{R}$ to refer to this set. When the domain and the range equal $\mathbb{R}$, you know that for any real number x, the resulting $f(x)$ will also be a real number.

When you find the domain, you are really watching for a few exceptions to the rules of math that can cause problems.

Remember these restrictions:

- You can't divide by 0.
- You can't find the square root of a negative number.

FINDING THE DOMAIN OF A FUNCTION

Example 1 What is the domain of $f(x) = x + 3$?

The domain is $\mathbb{R}$. You can use any real number for x.

Example 2 What is the domain of $f(x) = \frac{1}{x-2}$?

The domain is **all real numbers except for 2**; $x \neq 2$.
If x equals 2, the denominator of the fraction will be 0.

Example 3 What is the domain of $f(x) = \sqrt{x-2}$?

The domain is **all real numbers greater than or equal to 2**; $x \geq 2$.
If $x < 2$, then the number in the square root symbol will be negative.

When a number is excluded from the domain, the graph of the function will have an open circle to show that a value is missing.

Example 4 What is the domain and range of $g(x)$?

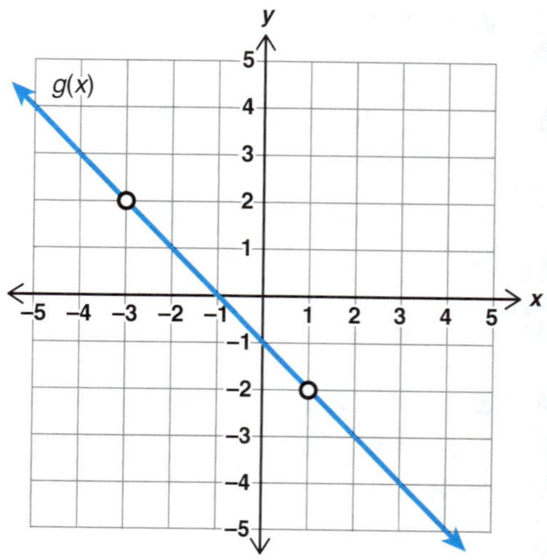

The domain is **all real numbers except 1 and –3**. If x is equal to 1 or –3, the graph has an open circle to show that there is no corresponding value for y.

The range is **all real numbers except 2 and –2**. There are no values of x that can produce either of these results. The open circles tell you that these values are missing from the range.

Find the domain of each function.

1. $f(x) = 3x - 2$ $\qquad g(x) = \sqrt{2x}$ $\qquad h(x) = \frac{2}{x+5}$

2. $f(x) = |\ x + 2\ |$ $\qquad g(x) = \frac{1}{4x}$ $\qquad h(x) = \sqrt{3x - 12}$

3. $f(x) = \frac{3x}{x}$ $\qquad g(x) = -\frac{1}{2}x + 2$ $\qquad h(x) = \sqrt{4x - x}$

Find the range of each function.

4. $f(x) = x^2$ $\qquad g(x) = |\ x^3\ |$ $\qquad h(x) = x - 1$

5. $f(x) = |\ 3x\ |$ $\qquad g(x) = x^3$ $\qquad h(x) = 4x + 7$

State the domain and range for each graphed function.

6.

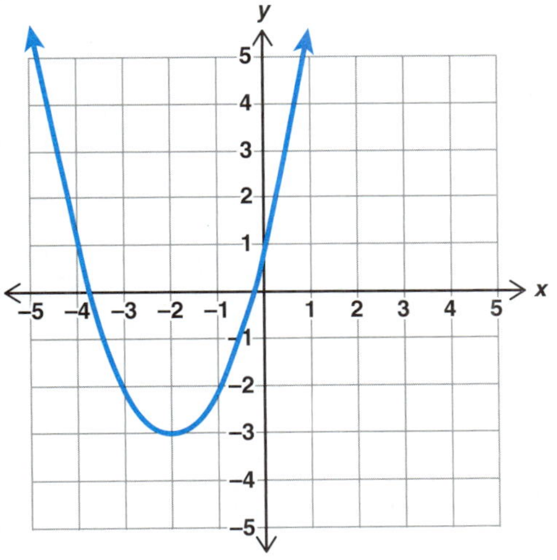

8.

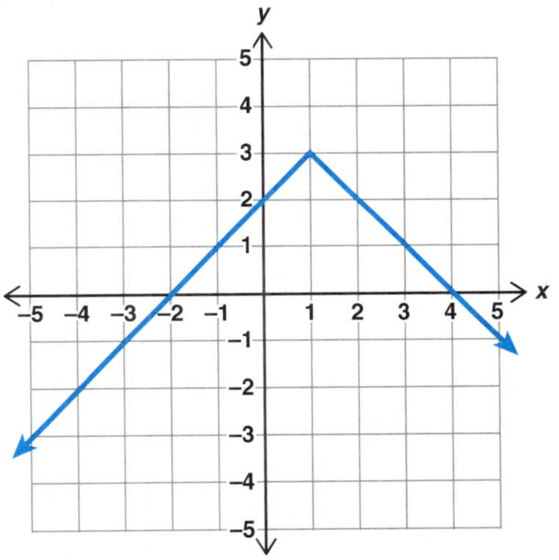

7.

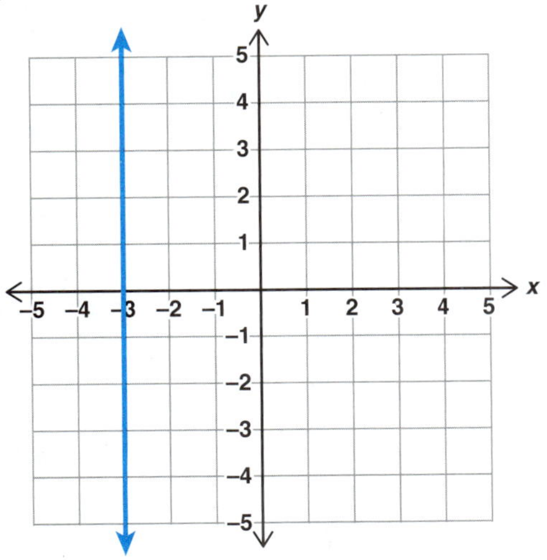

9.

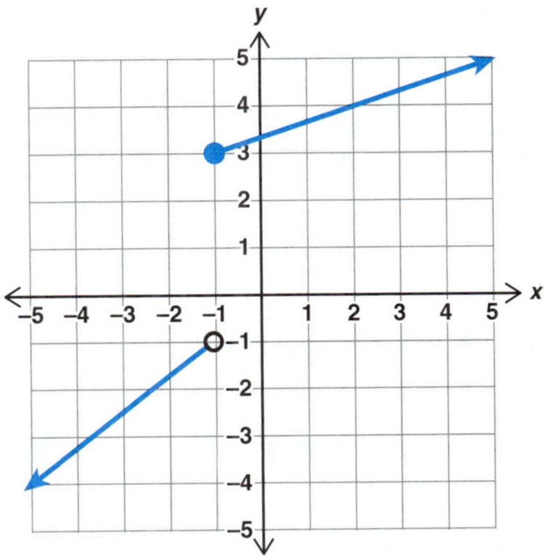

Answers start on page 177.

Problem Solver

COMPARING FUNCTIONS

On a test, you may be asked to compare functions and draw conclusions. Sometimes, two or more functions may be drawn on the same graph.

Example Tom, Alex, and Ron decide to run a 10-mile race. Data is collected at checkpoints along the course of the race. Using the data, each run can be expressed as a function and drawn on a graph. Which runner finished first?

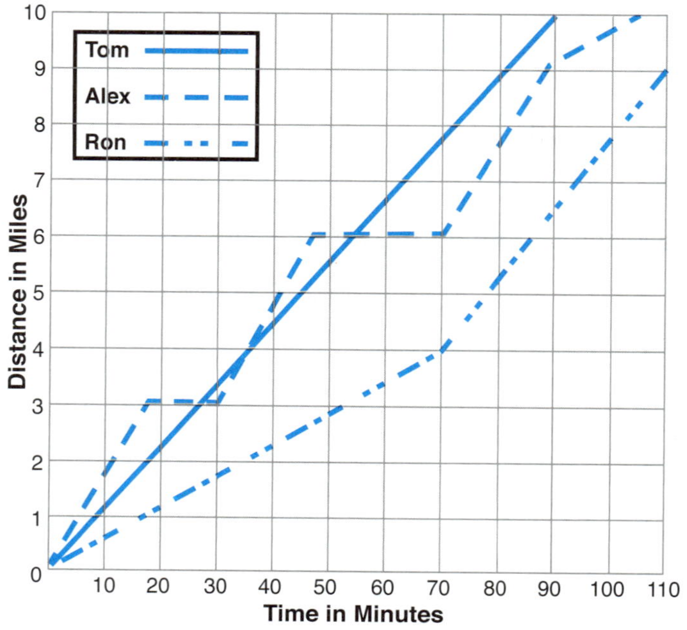

Examine the three lines. Who ran 10 miles in the least number of minutes?

Tom finished the 10-mile run in 90 minutes. At 90 minutes, Alex had run about 9 miles and Ron between 6 and 7 miles. **Tom won the race.**

Use the graph above to answer the questions.

1. Which runner was ahead after 3 miles?

2. Which runner ran faster at the end than he did at the beginning of the race? (***Hint:*** A change in the slope, or slant of a line, indicates a change in speed.)

3. Which person ran at a steady rate the entire race?

4. Which runner didn't finish the race?

5. If the race had ended at 5 miles, who would have won?

6. **Explain** Describe Alex's run in your own words. Describe the changes in his run. You may want to break his run into intervals.

You may be asked to draw conclusions about functions that are shown in different ways. Make sure you understand how the presentations are related before you answer the questions.

Example Elise wants to plant trees that will benefit the environment. She searches the Internet to find information about sugar maples and blue spruce trees. At a diameter of 8 inches, which type of tree is better at reducing CO_2 in the atmosphere?

Blue Spruce Facts

Annual Growth of Trunk Diameter: 0.5 inch

CO_2 Reduction

Diameter of Trunk	Annual CO_2 Reduction
2 in	28 lb
4 in	62 lb
6 in	106 lb
8 in	152 lb
10 in	203 lb

Annual Storm Water Recovery

Diameter of Trunk	Water Saved
2 in	123 gal
4 in	351 gal
6 in	723 gal
8 in	1,142 gal
10 in	1,626 gal

Sugar Maple Facts

The diameter of a sugar maple trunk grows at an average rate of 0.3 inch per year.

Its root system recovers storm water, which can be predicted using this function:

$$g(x) = 4x^2 + 104x - 123$$

where $g(x)$ represents gallons of water saved annually and x represents the diameter of the trunk in inches.

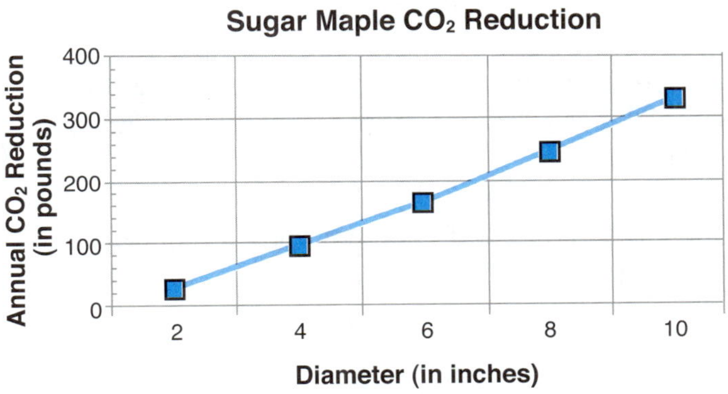

Read the first table. At 8 inches in diameter, a blue spruce removes 152 pounds of CO_2 from the atmosphere. Read the line graph. At 8 inches in diameter, a sugar maple removes over 200 pounds of CO_2. **The sugar maple is better at reducing CO_2.**

Use the information to answer the questions.

7. A blue spruce has a diameter of 14 inches. Based on the data, what is a good estimate for how many pounds of CO_2 the tree will reduce annually?

8. If both trees have a diameter of 8 inches, which saves more storm water?

9. At 6 inches in diameter, an apple tree removes 100 pounds of CO_2 and saves 150 gallons of storm water per year. How does an apple tree compare to a sugar maple and a blue spruce of the same diameter?

10. **Explain** A blue spruce and a sugar maple are planted, each with the same diameter of 2 inches. Ten years later, which tree will be better at storm water recovery? Show your work and explain your reasoning.

Answers start on page 178.

Problem Solver

FUNCTIONS IN THE REAL WORLD

In every profession, data is collected, analyzed, and displayed in tables and graphs. People then look for patterns in the data and use mathematical methods to write functions to model the data.

Example During an experiment, you measure the temperature of a liquid four times. You take the measurements in both Celsius and Fahrenheit. Your results are shown in the table to the right. Can you use the data to figure out how to convert any temperature from Celsius to Fahrenheit?

Celsius	Fahrenheit
5° C	41° F
20° C	68° F
25° C	77° F
35° C	95° F

Write the data from the table as ordered pairs.

(5, 41), (20, 68), (25, 77), (35, 95)

Plot the points on a graph. The points seem to form a line. Draw a line through the points.

You can use the point-slope form of a line to write the equation of the line. The equation is actually a function: $f(x) = 1.8x + 32$.

This function can be used to convert any temperature from Celsius to Fahrenheit.

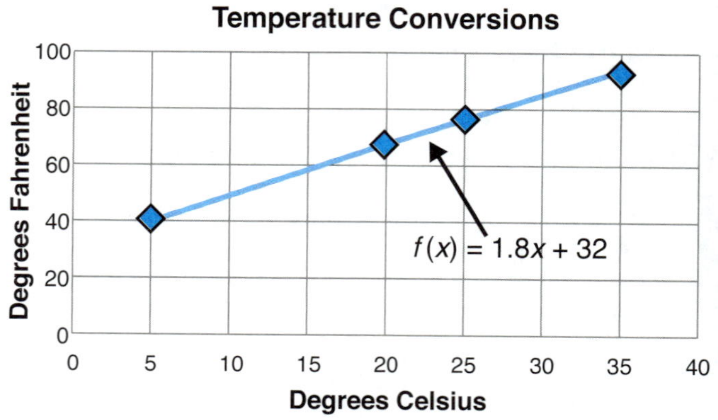

For example, 15°C is equal to how many degrees Fahrenheit?

Use the function and find $f(15)$.

$$\begin{aligned} f(x) &= 1.8x + 32 \\ f(15) &= 1.8 \cdot 15 + 32 \\ &= 27 + 32 \\ &= 59° \end{aligned}$$

The answer **59° Fahrenheit** is consistent with the graph above.

Solve.

1. Water boils at 100° Celsius. What is the boiling temperature of water in Fahrenheit?

2. Water freezes at 0° Celsius. Convert the temperature to Fahrenheit.

3. Surprisingly, when it is –40°, the temperature is the same on both the Celsius and Fahrenheit scales. Use the function to prove this statement is true.

On the GED® Math Test, you may be given a function that models a real situation. Make sure you understand what the function is telling you. Then substitute a value for x and solve for $f(x)$. Remember that on a graph, $f(x)$ is the same as the y-axis.

Example Did you know that you can estimate the air temperature by counting the number of times a common black cricket chirps? Dolbear's Law says that crickets chirp more frequently when it is warmer. After collecting data, Amos Dolbear wrote the function $T(n) = 50 + \frac{n-40}{4}$, where T is the temperature in degrees Fahrenheit and n is the number of chirps per minute.

Bill counts 80 chirps in 1 minute. What is the temperature in degrees Fahrenheit?

Use the function:

$$T(80) = 50 + \frac{80-40}{4}$$
$$= 50 + \frac{40}{4}$$
$$= 60°F$$

Remember to use the order of operations when you evaluate any function.

The temperature is **60 degrees Fahrenheit**.

Solve.

4. In baseball, the following function can be used to find the amount of time (T) in seconds that a batter has to react to a pitch thrown at x miles per hour.

$$T(x) = \frac{60.5}{1.47x}$$

a. A pitcher throws a fastball at 94 miles per hour. How many seconds does the batter have to react? Round to the nearest thousandth.

b. A pitcher throws a changeup at 76 miles per hour. How many seconds does the batter have to react? Round to the nearest thousandth.

5. Explain You can find the temperature in degrees Fahrenheit by counting the number of times a snowy tree cricket chirps in 13 seconds and adding 40. Refer to the example above. Which cricket chirps more quickly, the snowy tree cricket or the black cricket? How do you know?

6. Scientists model average global temperature (T) using this function:

$$T(x) = 0.02x + 50.5$$

where x is the parts per million of carbon dioxide concentration in the air.

a. Experts on climate change estimate that in 1960, the carbon dioxide concentration was about 320 parts per million. What was the average global temperature in 1960?

b. If the carbon dioxide concentration increases from 1960 rates by 200 parts per million, by how many degrees will the global temperature increase?

c. Explain In 2010, the carbon dioxide concentration was about 400 parts per million. Scientists warn that we need to reduce the concentration to 300 parts per million. What effect would this reduction have on global temperature?

Answers start on page 178.

Unit 4 Review

Solve as directed.

1. If you continue the pattern, what shape will be the 50th in the sequence?

2. Examine the series below.

−8, 5, 18, 31, 44, . . .

If −8 is the first number in the sequence, what number will be the tenth?

A. 109
B. 117
C. 122
D. 130

3. Assume 11 is the first term in the sequence. Which of the following rules can be used to find the *n*th number in the sequence below?

11, 20, 29, 38, 47, . . .

A. $-9n - 2$
B. $9n + 2$
C. $9n + 11$
D. $11n + 9$

4. Which of the following is a geometric series?

A. 8, 12, 16, 20, 24, . . .
B. 6, 4, 2, 0, −2, . . .
C. 5, 7, 9, 11, 13, . . .
D. 3, 6, 12, 24, 48, . . .

5. What is the next term in the series?

3, 15, 75, 375, ____________

Problems 6–10 refer to the functions below.

$f(x) = 5x - 9$

$g(x) = -2x^2 + 7$

$h(x) = |x^3 - x|$

6. Find $f(-2)$.

7. Find $g(3)$.

8. Find $g(-5)$.

9. Find $h(4)$.

10. Find $h(-4)$.

11. What is the domain of the function $f(x) = \frac{x-4}{x+5}$?

A. All real numbers
B. All real numbers except 4
C. All real numbers except −5
D. All real numbers except 4 and −5

Problems 12 and 13 refer to the graph of $f(x)$ below.

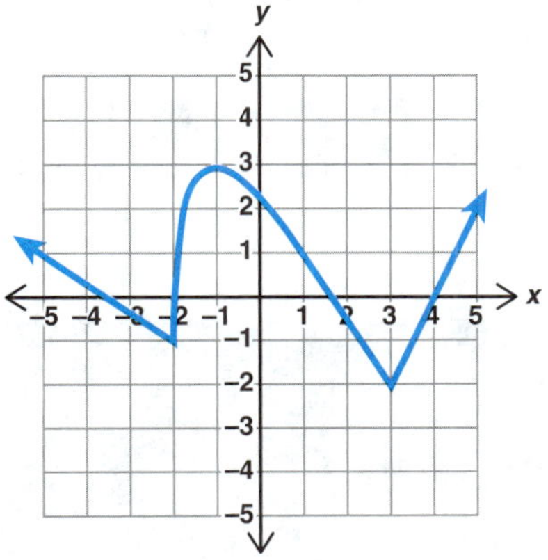

12. What is the value of x when $f(x)$ is at the highest point shown?

A. 5

B. 3

C. –1

D. –5

13. Which interval on the graph shows a decrease of 5 units?

A. $3 \le x \le 5$

B. $-1 \le x \le 3$

C. $-2 \le x \le -1$

D. $-5 \le x \le -2$

14. The domain of the function is all real numbers.

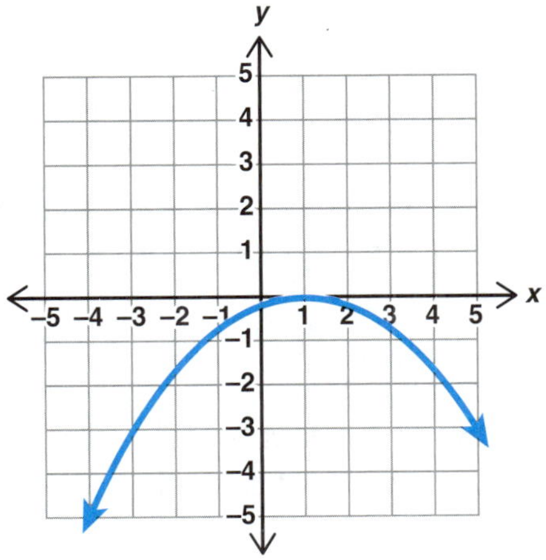

The range of the function is

A. All real numbers

B. All real numbers greater than 0

C. All real numbers less than –5

D. All real numbers less than or equal to 0

Does the graph represent a function? Select *Yes* or *No*.

15.

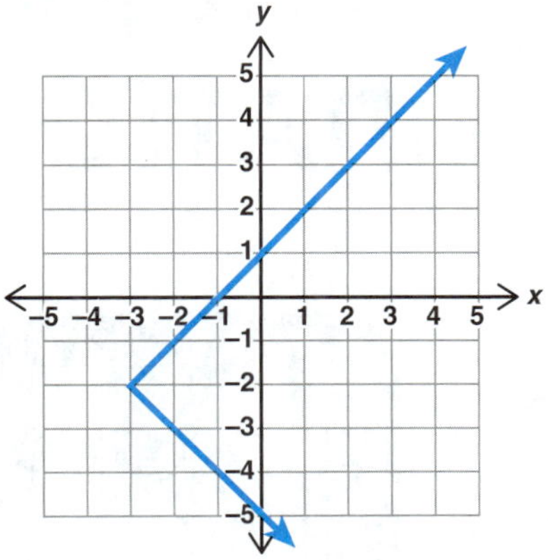

Yes

No

16.

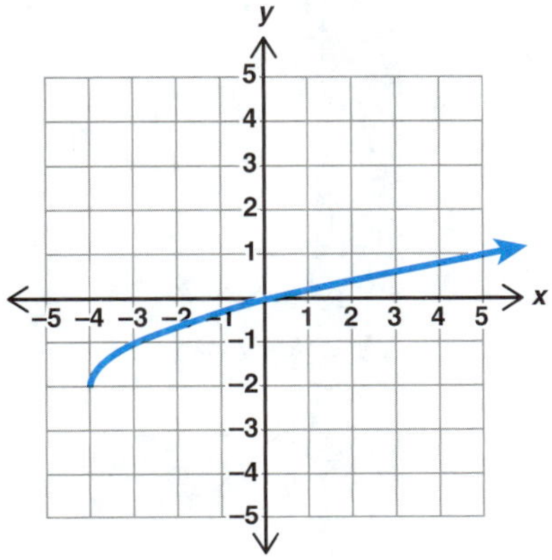

Yes

No

Problems 17 and 18 refer to this information.

Pediatricians model the average height (H) in inches of a boy who is m months old using the function $H(m) = 2.9\sqrt{m} + 20$.

17. At 48 months, Brad's son is 39 inches tall. Is Brad's son above or below average in height?

18. Matt's height (in inches) at various ages (in months) is recorded in the table.

Age (months)	12	18	24	30	36
Height (inches)	30	32	35	38	39

Describe Matt's growth pattern. How does it compare to average?

Answers start on page 179.

UNIT

5

POLYNOMIALS AND RATIONAL EXPRESSIONS

SKILLS

- Simplifying polynomial expressions
- Multiplying polynomial expressions
- Factoring polynomials
- Performing operations with rational expressions
- Simplifying rational expressions
- Evaluating rational expressions

TOOLS AND STRATEGIES

- Use FOIL to multiply binomials
- Find undefined values of rational expressions

WHEN YOU LEARNED the basics of algebra, you learned how to write mathematical expressions using variables. You also learned how to use the rules of exponents.

Polynomials combine these two skills. In this unit, you will learn how to work with expressions that involve terms with exponents. You will also learn how to write these expressions as a product, or a multiplication problem, with two factors.

Rational expressions put algebraic expressions in the numerators and denominators of fractions. Rational expressions are like ratios: they compare two expressions. The rules of fractions still apply to rational expressions. In this unit, you will learn how to use your knowledge of fractions to work with rational expressions.

USING POLYNOMIALS AND RATIONAL EXPRESSIONS

Polynomials and rational expressions are used:

- In geometric formulas for perimeter, area, volume, and surface area
- When writing computer programs to protect data
- To measure distances in outer space
- When analyzing and predicting changes in stock prices
- When searching for the best home loan

Read the following items. Describe some of your experiences in writing or discuss them with a partner.

1. Have you ever had car trouble? When you take your car in for repairs, the mechanic will give you a quote, or an estimate, for the cost of the repairs. What factors do you think affect how much it will cost to fix a car?

2. When a teacher asks you to simplify a fraction, what does that mean? How do you know when a fraction is completely simplified?

3. When you purchase car insurance, the insurance company decides how much to charge you based on whether it thinks you will be a good driver. What factors do you think insurance companies look at?

4. One of the most important properties in algebra is the distributive property. Describe how to use the distributive property. In what kinds of equations do you use the distributive property?

MATH TALK

The formulas for the volumes of three-dimensional objects use polynomials. You already know how to find the volumes of standard shapes such as cubes and rectangular prisms. How would you find the volume of a more unusual shape?

Picture the trunk of a car. How would you find its volume? Could you divide it into shapes that you are more familiar with?

Discuss the process you would use to find the volume of a car's trunk with a partner. Then think about a way to write down your process using mathematical symbols.

SIMPLIFY POLYNOMIAL EXPRESSIONS

All algebraic expressions are made up of **terms**. A term can be a variable, a number (also called a **constant**), or one or more variables multiplied by a number.

You have already worked with terms. Here are some examples:

$$x \qquad 3y \qquad -5 \qquad a^2b \qquad \frac{w}{8}$$

Like terms have exactly the same variables and exponents. The terms x^3 and $4x^3$ are like terms because they have the same variable raised to the same power. **Unlike terms** have different variables or the same variable raised to different powers.

When a variable is multiplied by a number, the number is called a **coefficient**. The coefficient of $5ab$ is 5. Like terms can have different coefficients.

A **polynomial** is an expression with one term or the sum or difference of unlike terms. Polynomials are classified by the number of terms they contain. A **monomial** has one term. The examples at the top of the page are all monomials.

A **binomial** is the sum or difference of two unlike terms. The following are binomials:

$$x + 8y \qquad 6y^2 - 1 \qquad a^3 + 5ab$$

A **trinomial** is the sum or difference of three unlike terms. The following are trinomials:

$$x^2 + 5x + 6 \qquad 9x^3 - 7y + z^2 \qquad 12a^2 - 17ab - 7b^2$$

Classify each expression as a monomial, binomial, or trinomial.

1. $-2x^3$

2. $x^2 + 8x + 12$

3. $5a^5b^2$

4. $9 - x^3 + y^2$

5. $k + m$

6. $y^2 - 2y$

7. $z^3 - xz + x^2$

8. -91

9. $9n - 80$

10. $7mn$

To simplify polynomials, combine like terms.

SIMPLIFYING POLYNOMIALS

Example Simplify: $5x^2 - 3x + 4 + 6x - 9$

Step 1
Group like terms.

$5x^2 - 3x + 6x + 4 - 9$

Step 2
Combine like terms by adding the coefficients.

$5x^2 - 3x + 6x + 4 - 9$
$= 5x^2 + 3x - 5$

Step 3
Write the simplified expression.

$5x^2 + 3x - 5$

Simplify the following expressions.

11. $-4r^3 - 4 - 7r^3 + 3$

12. $7x - 1 - 6x^2 + 8$

13. $4k + 6 - 6k + 2k$

14. $-1 - 2y^2 + y^3 + 4 + 3y^2$

15. $-6n - 7n - 8 + 6n^2 + 4n$

16. $2ab^2 - 6b - 8 + 3ab^2 - 2ab$

When you subtract polynomials, you will need to distribute the minus sign to the terms in parentheses that follow.

ADDING AND SUBTRACTING POLYNOMIALS

Example Simplify: $(y^2 - 2y) - (2y^2 + 5y)$

Step 1
Use the distributive property.

$(y^2 - 2y) - (2y^2 + 5y)$
$= y^2 - 2y - 2y^2 - 5y)$

Step 2
Group like terms.

$y^2 - 2y^2 - 2y - 5y$

Step 3
Combine like terms.

$-y^2 - 7y$

Simplify the following expressions.

17. $(4m - 2) + (-8m + 7m^2)$

18. $(-3 - 4n) - (-5 + 6n)$

19. $(-p^2 - p) + (7 + 5p^2)$

20. $(8x^3 + 8x^2 + 1) - (-5x^3 - 2x^2)$

21. $(3a - 8a^2 + 6) - (2a^2 + 6a + 7)$

22. $(-2x - 8xy^2) - (6xy^2 - 2y - 7x)$

Answers start on page 179.

MULTIPLY AN EXPRESSION BY A MONOMIAL

A monomial can have two parts: the coefficient and the variables. When you multiply two monomials, multiply each part separately. First multiply the coefficients. Then multiply each variable.

Use the rules of exponents to multiply variables. Remember, when you multiply variables with the same bases, keep the base the same and add the exponents. For a complete review of the rules of exponents, see *Math Sense 2: Focus on Problem Solving,* pp. 48–49.

MULTIPLYING MONOMIALS

Example Multiply: $2x^3y \cdot 7xy^2$

Step 1
Multiply the coefficients.

$2x^3y \cdot 7xy^2$
$2 \cdot 7 = 14$

Step 2
Multiply the variables.

$x^3 \cdot x = x^{3+1} = x^4$
$y \cdot y^2 = y^{1+2} = y^3$

Step 3
Write the simplified expression.

$14x^4y^3$

You can also multiply an expression by a monomial. Use the distributive property to multiply each term in the expression by the monomial.

MULTIPLYING AN EXPRESSION BY A MONOMIAL

Example 1 Multiply: $7(z^3 + 5)$

Step 1
Use the distributive property.

$7(z^3 + 5)$
$(7 \cdot z^3) + (7 \cdot 5)$

Step 2
Multiply each pair of monomials.

$7 \cdot z^3 = 7z^3$
$7 \cdot 5 = 35$

Step 3
Write the expression.

$7z^3 + 35$

Example 2 Multiply: $-3ab(4a^2 + ab)$

Step 1
Use the distributive property.

$-3ab(4a^2 + ab)$
$(-3ab \cdot 4a^2) + (-3ab \cdot ab)$

Step 2
Multiply.

$-3ab \cdot 4a^2 = -12a^3b$
$-3ab \cdot ab = -3a^2b^2$

Step 3
Write the expression.

$-12a^3b - 3a^2b^2$

Multiply.

1. $2x \cdot x^2$ $\quad$ $4ab \cdot 8b$ $\quad$ $-10m^2n \cdot mn^3$

2. $3n(6n+8)$ $\quad$ $3(-7b+2)$ $\quad$ $-5y(-2x-8)$

3. $8v(-8v^2-2v)$ $\quad$ $7m^2(-4m-1)$ $\quad$ $-(-3x^3+4x^2)$

4. $-m(-k+5km)$ $\quad$ $-5xy(-6x-y)$ $\quad$ $-4a^2b(-a^2-8ab)$

5. $3(4p^2-p+5)$ $\quad$ $\frac{1}{2}(-6n^2+8n+6)$ $\quad$ $8x(2x^2+2x-5)$

6. $-(a^2-6a-5)$ $\quad$ $-6v(-3v^2-6v-4)$ $\quad$ $xy(x^2+xy-y^3)$

CORE CONNECTIONS: Modeling with Polynomial Functions

Mathematicians use functions to represent or estimate real-life situations. This is called **modeling**. Many of these models use polynomial expressions.

Example When a driver slams on the brakes, the car doesn't stop immediately. The distance required to stop a car depends on how fast the car is traveling. This situation is modeled by the function $f(x) = 2.2x + \frac{1}{20}x^2$, where x is the car's velocity and $f(x)$ is the distance in feet needed to stop the car. If a car is traveling at a speed of 35 mph, how many feet will it need to stop?

Substitute 35 for x in the function.

$$f(35) = 2.2(35) + \frac{1}{20}(35)^2 = 138.25 \text{ ft}$$

A car traveling 35 mph would take **138.25 feet** to stop.

Solve as directed.

1. Use the function above to find the distance required to stop a car traveling at a rate of:

a. 50 mph

b. 75 mph

2. A clothing company models the cost of producing T-shirts with the function $C = n(30 - 0.02n) + 200$, where n is the number of T-shirts produced and C is the total cost. How much would it cost to produce 500 T-shirts?

Answers start on page 180.

MULTIPLY BINOMIALS USING FOIL

You have learned how to use the distributive property to multiply each term inside parentheses by the monomial outside parentheses.

$$4x(x^2 - 6)$$

You can multiply two binomials using the same idea. Each term in the second binomial must be multiplied by each term in the first binomial. A method called **FOIL** helps you keep track of the multiplication. FOIL stands for First, Outer, Inner, Last.

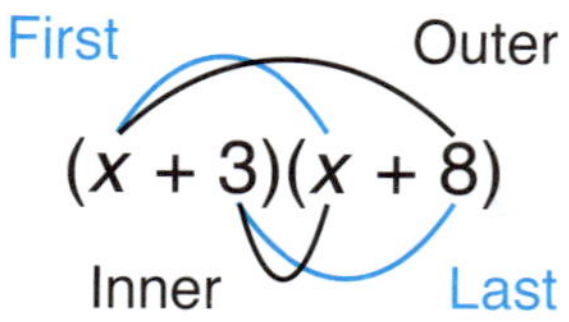

Multiply the terms in this order, and then combine like terms to simplify the expression.

USING FOIL TO MULTIPLY BINOMIALS

Example 1 Multiply: $(x + 3)(x + 8)$

Step 1
Use FOIL to multiply.

First: $x \cdot x = x^2$
Outer: $x \cdot 8 = 8x$
Inner: $3 \cdot x = 3x$
Last: $3 \cdot 8 = 24$

Step 2
Write the expression.

$x^2 + 8x + 3x + 24$

Step 3
Combine like terms.

$x^2 + 11x + 24$

$(x + 3)(x + 8) = \mathbf{x^2 + 11x + 24}$

Example 2 Multiply: $(2y - 1)(y - 6)$

Step 1
Use FOIL to multiply.

First: $2y \cdot y = 2y^2$
Outer: $2y \cdot -6 = -12y$
Inner: $-1 \cdot y = -y$
Last: $-1 \cdot -6 = 6$

Step 2
Write the expression.

$2y^2 - 12y - y + 6$

Step 3
Combine like terms.

$2y^2 - 13y + 6$

$(2y - 1)(y - 6) = \mathbf{2y^2 - 13y + 6}$

Use the FOIL method to multiply.

1. $(b + 7)(b - 1)$	$(r + 8)(r - 2)$	$(k + 1)(k + 5)$
2. $(m - 2)(m + 1)$	$(x - 3)(x - 6)$	$(p + 6)(p + 4)$
3. $(n + 12)(n + 7)$	$(b - 10)(b + 4)$	$(3a + 3)(a - 7)$
4. $(7x + 2)(x - 5)$	$(2n + 9)(2n + 3)$	$(5p - 1)(2p + 5)$
5. $(v - 4)(6v + 5)$	$(-a + 1)(9a - 8)$	$(m + 4)(m + 4)$
6. $(n + 5)(n - 5)$	$(2x + 3)(2x + 3)$	$(4k - 1)(4k + 1)$

Solve as directed.

7. Which of the following expressions could represent the area of the rectangle?

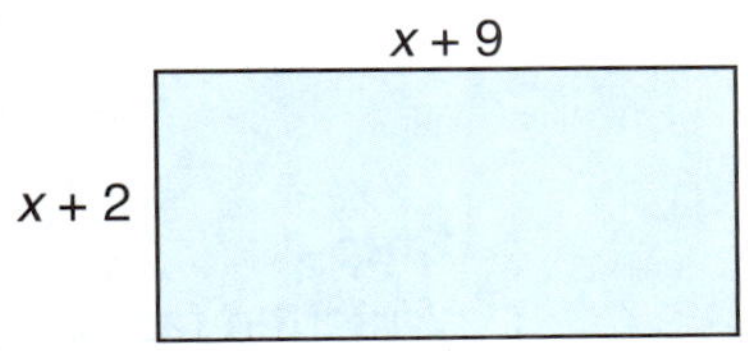

A. $2x + 11$
B. $x^2 + 18$
C. $x^2 + 2x + 18$
D. $x^2 + 11x + 18$

8. Write and simplify an expression to represent the area of the triangle below.

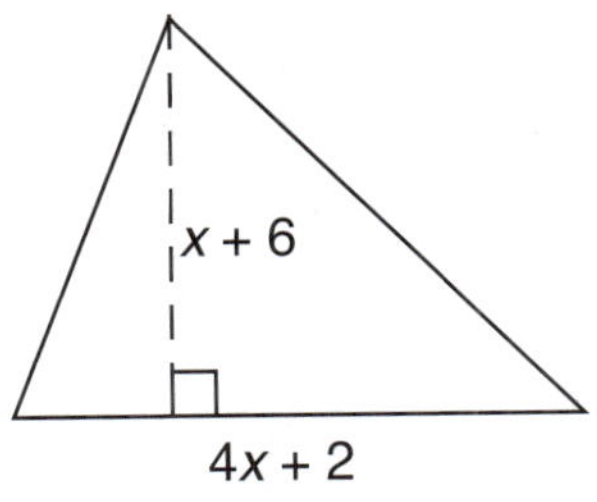

9. Write and simplify an expression to represent the following:

The sum of a number and 10 multiplied by the difference of the same number and 3.

10. Explain Steve does the following work for a problem.

$(x + 8)(x - 8)$

$x \cdot x = x^2$
$x \cdot 8 = 8x$
$8 \cdot -8 = -64$

Answer: $x^2 + 8x - 64$

Michelle points out that Steve made a mistake. What error did Steve make? What should the answer be?

Answers start on page 180.

FACTOR OUT A MONOMIAL

When finding the factors of a number, you look for values that multiply to make that number. For example, 3 and 5 are factors of 15 because $3 \times 5 = 15$.

Factoring a polynomial expression is the opposite of using the distributive property. When factoring a polynomial, you look for the greatest common factor (GCF) of all the terms in the expression. You then divide by, or factor out, the greatest common factor.

FACTORING OUT A MONOMIAL

Example 1 Factor the expression $6x^2 + 18x + 12$.

Step 1
Find the greatest common factor of the coefficients.

The GCF of 6, 18, and 12 is 6.

Step 2
Find the greatest common factor of the variables.

The third term has no variable, so there is no common variable factor.

Step 3
Divide each term by the GCF.

The GCF of the expression is 6.

$$\frac{6x^2}{6} + \frac{18x}{6} + \frac{12}{6} = x^2 + 3x + 2$$

Step 4
Write the factors as a product.

$6(x^2 + 3x + 2)$

Example 2 Factor the expression $24x^4 + 48x^3 + 56x^2$.

Step 1
Find the greatest common factor of the coefficients.

The GCF of 24, 48, and 56 is 8.

Step 2
Find the greatest common factor of the variables.

What do x^4, x^3, and x^2 have in common?

Each term has at least an x^2.

Step 3
Divide each term by the GCF.

The GCF of the expression is $8x^2$.

$$\frac{24x^4}{8x^2} + \frac{48x^3}{8x^2} + \frac{56x^2}{8x^2} = 3x^2 + 6x + 7$$

Step 4
Write the factors as a product.

$8x^2(3x^2 + 6x + 7)$

You can always check your work by multiplying using the distributive property. You should get the original expression.

The greatest common factor of a polynomial expression is a monomial. This kind of factoring is called *factoring out a monomial* because you pull the greatest common factor to the outside of the expression.

Factor each expression.

1. $3x^2 + 33$	$15y^3 - 35y$	$-36m^3 + 16m$
2. $25n^2 + 40n + 45$	$40 - 16k - 8k^2$	$-35x^3 + 42x^2 - 56x$
3. $10n^3 + 30n^2 - 10n$	$8x^4 + 16x^3 + 14x^2$	$-63v^4 - 36v^2 + 45v$
4. $90ab^2 - 35ab - 15a$	$24x^2 - 56xy - 16x$	$10m^2n^2 - 5m^2 + 30m$
5. $10a^2b + 20ab^2 + 8ab$	$-49x^3y + 21xy^2 + 21xy$	$12 - 27st + 18t^2$
6. $-y^3 - yx^2 - yx$	$40m^2 + 64n^2 - 72n$	$15u^3v + 9u^2v^2 + 9u^2v$

Solve as directed.

7. Which monomial is the greatest common factor of the terms $18m^3np$, $-108mn^2p$, and $54m^2n^2p$?

A. $3m^2n^2p$

B. $18mnp$

C. $18m^3n^2p$

D. $54m^6n^5p^3$

8. In a rectangle, the width can be represented by the expression $8a$. The area of the rectangle is $32a^2b + 88ab + 24a$. Which of the following expressions could represent the length of the rectangle?

A. $4ab + 11b + 3$

B. $4a^2b + 11ab + 3a$

C. $32ab + 88b + 24$

D. $256a^3b + 704a^2b + 192a^2$

9. The formula for the surface area of a cylinder is $SA = 2\pi r^2 + 2\pi rh$. Which of the following correctly factors the formula?

A. $SA = h(2\pi r^2 + 2\pi r)$

B. $SA = 2\pi r(r^2 + rh)$

C. $SA = 2\pi r(r + h)$

D. $SA = 2\pi rh(2\pi r^2)$

10. Explain Ginnie factored the expression $65x^2y + 39xy^2 - 26xy$. She says the correct factorization is $13(5x^2y + 3xy^2 - 2xy)$ because if she uses the distributive property, she gets the original expression. Is Ginnie's answer correct? If not, what is the correct factorization? Explain your thinking.

Answers start on page 180.

USE FOIL TO FACTOR TRINOMIALS

You use FOIL to multiply two binomial factors. The answer is a trinomial that looks like this: $ax^2 + bx + c$, where a, b, and c are numbers. You can also use the idea behind FOIL to help you find the binomial factors that make a trinomial.

Look at the diagram to the right. Can you see how the terms in the binomials are used to find the product?

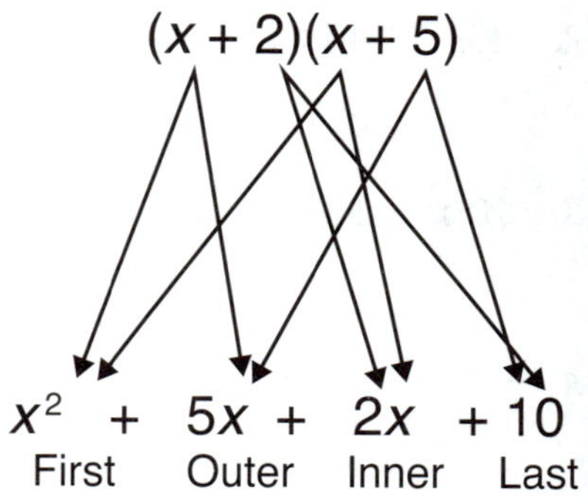

The simplified trinomial is $x^2 + 7x + 10$. When you add 5 and 2, you get the coefficient of the middle term, $7x$. You multiplied 5 and 2 to get 10, the last term.

When you factor a trinomial, look for a pair of numbers that multiply to make the last term and add to make the middle term.

FACTORING A TRINOMIAL

Example Factor: $x^2 + 4x - 12$

Step 1
Find the factors of x^2.

$x \cdot x = x^2$

Step 2
List the factors of –12.

1 and –12	–1 and 12
2 and –6	–2 and 6
3 and –4	–3 and 4

Step 3
Look for the factor pair that combines to make 4.

$-2 + 6 = 4$

Step 4
Write the factors.

$(x - 2)(x + 6)$

The factorization of $x^2 + 4x - 12$ is $\mathbf{(x - 2)(x + 6)}$.

Factor the following trinomials.

1. $x^2 + 8x + 15$	$x^2 - 3x - 28$	$x^2 - 12x + 20$
2. $x^2 + 4x - 32$	$x^2 + 14x + 45$	$x^2 - 14x + 48$
3. $x^2 - 4x + 3$	$x^2 - x - 6$	$x^2 + 17x + 60$
4. $x^2 + 21x + 80$	$x^2 - x - 56$	$x^2 + 3x - 108$

When you factor a trinomial of the form $ax^2 + bx + c$ when $a > 1$, you will have to use trial and error.

FACTORING TRINOMIALS WHEN $a > 1$

Example Factor: $2x^2 + 9x + 4$

Step 1	Step 2	Step 3	Step 4
Find the factors of $2x^2$.	List the factors of 4.	Use trial and error to find the correct factor pair.	Write the factors.
$2x \cdot x = 2x^2$	4 and 1 2 and 2	$(2x + 4)(x + 1) = 2x^2 + 6x + 4$ $(2x + 1)(x + 4) = 2x^2 + 9x + 4$ $(2x + 2)(x + 2) = 2x^2 + 6x + 4$	$(2x + 1)(x + 4)$

The factorization of $2x^2 + 9x + 4$ is $\mathbf{(2x + 1)(x + 4)}$.

Factor the following trinomials.

5. $2x^2 - 11x - 21$ $\quad 2x^2 - 9x + 7$ $\quad 3x^2 + x - 10$

6. $3x^2 + 16x - 12$ $\quad 5x^2 + 14x + 8$ $\quad 4x^2 - 24x + 27$

Some polynomials follow specific patterns. It is a good idea to memorize these patterns to make factoring easier.

When you subtract two perfect squares, it is called a **difference of squares**.

$a^2 - b^2 = (a + b)(a - b)$

Example Factor: $x^2 - 49$

The square root of x^2 is x.
The square root of 49 is 7.

$x^2 - 49 = (x + 7)(x - 7)$

Trinomial squares have two identical factors. These trinomials are perfect squares.

$x^2 + 2ax + a^2 = (x + a)(x + a)$ or $(x + a)^2$
$x^2 - 2ax + a^2 = (x - a)(x - a)$ or $(x - a)^2$

Example Factor: $x^2 + 6x + 9$

The square root of 9 is 3, and 3(2) = 6. This fits the pattern of a trinomial square.

$x^2 + 6x + 9 = (x + 3)(x + 3)$ or $(x + 3)^2$

Factor the following trinomials.

7. $x^2 - 100$ $\quad x^2 - 4x + 4$ $\quad 4x^2 - 9$

8. $x^2 - 18x + 81$ $\quad x^2 + 22x + 121$ $\quad x^2 - 1$

Answers start on page 180.

Mixed Review

Simplify the following polynomial expressions.

1. $7n^2 + 3 - 2n^2$ $\qquad$ $-5b - 8b^2 + 10b^2 - 4b$

2. $2x^2 - 3x + 7 - 7x$ $\qquad$ $-r^2 - 3 - 6r^2 + 7r$

3. $2p^3 - 6p^2 + p - 8p^2 - 2p^3$ $\qquad$ $-3 - 3n + 3n^2 - 7n + 9$

4. $(5m^2 - 9) + (3 - m^2)$ $\qquad$ $(y - 7) - (y^2 + 6y)$

5. $(5x^2 + x - 4) + (-3x^2 + 8)$ $\qquad$ $(-8r^2 - 6r + 3) - (-3r^2 + 2r)$

6. $(-5a^3 - a^2) + (-5a^2 - a - 1)$ $\qquad$ $(5v^3 + 4v^2 + 6) - (-7v^3 - 8v^2 + 10)$

Multiply.

7. $4u^2 \cdot -3v^2$ $\qquad$ $7xy \cdot x^2$ $\qquad$ $5m^2 \cdot -2n$

8. $2a(5a - 1)$ $\qquad$ $6u^2(9 - 6u)$ $\qquad$ $7y(x - 7y)$

9. $8(2n^2 + 8n - 3)$ $\qquad$ $4(7a^2 - 8a + 1)$ $\qquad$ $6x(x^2 - 5x - 4)$

10. $-3p(p^2 - 6p - 1)$ $\qquad$ $4xy(7x - 4y^2)$ $\qquad$ $5b^2(-a^2 + ab)$

11. $(n - 5)(n + 6)$ $\qquad$ $(x + 11)(x + 3)$ $\qquad$ $(y - 18)(y - 2)$

12. $(2x - 3)(x - 3)$ $\qquad$ $(3m - 8)(m - 6)$ $\qquad$ $(4p + 1)(p - 7)$

13. $(b - 8)(b + 8)$ $\qquad$ $(k + 7)(k + 7)$ $\qquad$ $(2x - 1)(2x - 1)$

Factor.

14. $k^2 + 6k$ $2m^2 - 26m$ $10uv + 15u^2$

15. $9x + 27y - 63$ $m^2n + mn^2$ $-12ab^2 - 48b^2$

16. $34k^3m - 17k$ $4a^3 - 40a^2 + 32a$ $-33m^4 - 77m^2 + 99m$

17. $8x^2y + 28xy^2 - 20xy$ $uv^3 - u^2v^2 + uv^2$ $80k^3 + 50kj - 100k$

18. $x^2 - 9x + 14$ $v^2 + 6v - 7$ $p^2 - 19p + 90$

19. $b^2 - 4b - 21$ $x^2 + 17x + 52$ $m^2 - 7m - 60$

20. $2n^2 + 11n + 15$ $3y^2 - 5y - 2$ $5k^2 + 37k + 14$

21. $b^2 + 20b + 100$ $x^2 - 121$ $v^2 - 6v + 9$

Solve as directed.

22. Write an expression to represent the area of the square below.

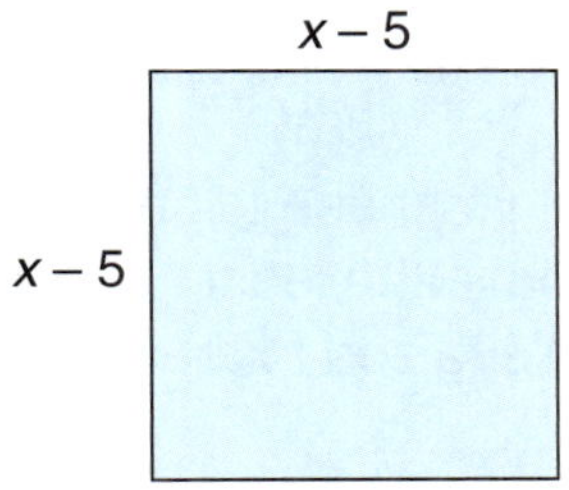

23. Silas is factoring $10x^2 - 41x - 18$. He decides that the first terms are $5x$ and $2x$, as shown below.

$(5x + \_\_\_\_\_)(2x - \_\_\_\_\_)$

Fill in the blanks and complete the factorization.

24. Write an expression to represent the perimeter of the triangle below.

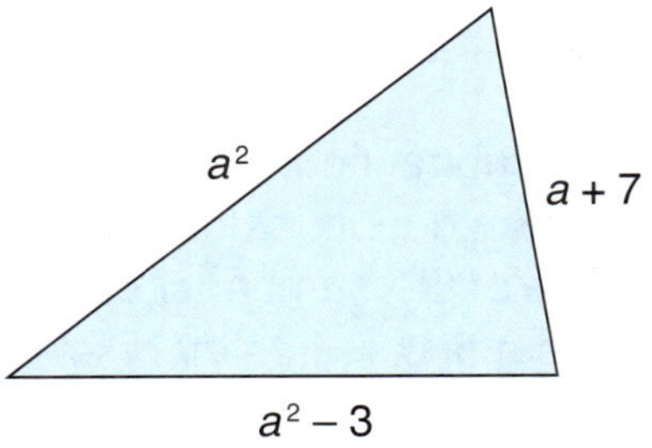

25. Explain Explain the error a student made when simplifying the following expression. What should the answer be?

$$2x^2y + 5y - 7xy^2 + 8y$$
$$= 2x^2y - 7xy^2 + 5y + 8y$$
$$= -5x^2y^2 + 13y$$

Answers start on page 181.

ADD AND SUBTRACT RATIONAL EXPRESSIONS

A ratio compares two numbers and can be written with words, with a colon, or as a fraction. For example, the ratio "3 to 5" can be written as 3:5 or $\frac{3}{5}$.

A **rational expression** is a ratio that compares two polynomials. Rational expressions are also written as fractions. Both the numerator and the denominator of a rational expression must be polynomials. Rational expressions can't have fractional exponents, negative exponents, or radicals.

The rules that apply to working with fractions apply to rational expressions. When you add or subtract fractions with like denominators, add or subtract the numerators and keep the denominator the same. You can add rational expressions the same way.

RATIONAL EXPRESSIONS WITH LIKE DENOMINATORS

Example Add: $\frac{4m^2}{n^3} + \frac{m^2 - 3m}{n^3}$

Step 1

The denominators are the same. Add the numerators.

$$\frac{4m^2}{n^3} + \frac{m^2 - 3m}{n^3} = \frac{4m^2 + m^2 - 3m}{n^3}$$

Step 2

Simplify the numerator.

$$\frac{4m^2 + m^2 - 3m}{n^3} = \frac{5m^2 - 3m}{n^3}$$

When adding or subtracting rational expressions with unlike denominators, you need to find the lowest common denominator. The lowest common denominator must have the least common multiple of the number coefficients. It must also have each variable raised to the highest power that occurs in either denominator.

RATIONAL EXPRESSIONS WITH UNLIKE DENOMINATORS

Example Subtract: $\frac{2}{3xy} - \frac{x + 3}{5x^2y}$

Step 1

Find the lowest common denominator.

The least common multiple of 3 and 5 is 15.

The highest power of x is x^2.

The highest power of y is y.

The lowest common denominator is $15x^2y$.

Step 2

Rewrite both expressions using the lowest common denominator.

$$\frac{2}{3xy} \cdot \frac{5x}{5x} = \frac{10x}{15x^2y}$$

$$\frac{x + 3}{5x^2y} \cdot \frac{3}{3} = \frac{3(x + 3)}{15x^2y} = \frac{3x + 9}{15x^2y}$$

Step 3

Subtract the expressions.

$$\frac{10x}{15x^2y} - \frac{3x + 9}{15x^2y}$$

$$= \frac{10x - (3x + 9)}{15x^2y}$$

$$= \frac{10x - 3x - 9}{15x^2y}$$

$$= \frac{7x - 9}{15x^2y}$$

Some rational expressions have a binomial in the denominator. To add these rational expressions, simply multiply the binomials to find the common denominator.

RATIONAL EXPRESSIONS WITH BINOMIAL DENOMINATORS

Example Add: $\frac{7x}{x+4} + \frac{8}{x-3}$

Step 1
Find a common denominator by multiplying the denominators.

The lowest common denominator is $(x + 4)(x - 3)$.

Step 2
Rewrite the expressions using the common denominator.

$$\frac{7x}{x+4} \cdot \frac{x-3}{x-3} = \frac{7x(x-3)}{(x+4)(x-3)} = \frac{7x^2-21x}{(x+4)(x-3)}$$

$$\frac{8}{x-3} \cdot \frac{x+4}{x+4} = \frac{8(x+4)}{(x+4)(x-3)} = \frac{8x+32}{(x+4)(x-3)}$$

Step 3
Add and simplify. Leave the denominator in factored form.

$$\frac{7x^2-21x}{(x+4)(x-3)} + \frac{8x+32}{(x+4)(x-3)}$$

$$= \frac{7x^2-21x+8x+32}{(x+4)(x-3)}$$

$$= \frac{7x^2-13x+32}{(x+4)(x-3)}$$

Add or subtract as indicated.

1. $\frac{a-4b}{10ab} + \frac{a+b}{10ab}$ $\quad$ $\frac{5m+n}{8n} - \frac{m-n}{8n}$ $\quad$ $\frac{u+3v}{24u^3} + \frac{u-2v}{24u^3}$

2. $\frac{6m}{4} + \frac{5m}{3n}$ $\quad$ $\frac{2u}{5} - \frac{3u}{2}$ $\quad$ $\frac{4}{5} - \frac{5x}{6x^2}$

3. $\frac{x+6y}{3} - \frac{3}{4xy}$ $\quad$ $\frac{5n}{6m} + \frac{m-8n}{3n}$ $\quad$ $\frac{8}{3ab} + \frac{3a+b}{4b^2}$

4. $\frac{x+3}{12x^3-12x^2} - \frac{x-10}{12x^3-12x^2}$ $\quad$ $\frac{6}{3m-15} - \frac{2m}{3m-15}$ $\quad$ $\frac{r+2}{r^2+9r+20} + \frac{r+5}{r^2+9r+20}$

5. $\frac{4p}{p+2} - \frac{3}{p+4}$ $\quad$ $\frac{5k}{k-5} - \frac{3k}{5}$ $\quad$ $\frac{3}{3x-5} - \frac{4x}{x+4}$

6. $\frac{n-1}{n+1} - \frac{n+2}{4}$ $\quad$ $\frac{a-5}{a-1} + \frac{a}{a+7}$ $\quad$ $\frac{n+2}{n+3} + \frac{4}{n+6}$

Solve as directed.

7. A rational expression has no negative exponents. Rewrite x^{-2} as a rational expression.

8. **Explain** Add: $\frac{1}{2y} + 3y$
How can you write $3y$ as a rational expression? (***Hint:*** Think about how you would write the number 4 as a fraction.)

Answers start on page 181.

MULTIPLY RATIONAL EXPRESSIONS

The rules for multiplying rational expressions are the same as the rules for multiplying fractions. First, look for ways to cancel. Then multiply the numerators and the denominators straight across.

MULTIPLYING RATIONAL EXPRESSIONS WITH MONOMIALS

Example Multiply: $\frac{6k^2}{5} \cdot \frac{4m^2}{18km}$

Step 1
Cancel. Remember to look for common factors in the coefficients and the variables.

$$\frac{\overset{k}{\cancel{6k^2}}}{5} \cdot \frac{\overset{m}{4\cancel{m^2}}}{\underset{3}{\cancel{18km}}}$$

Step 2
Multiply straight across.

$$\frac{\overset{k}{\cancel{6k^2}}}{5} \cdot \frac{\overset{m}{4\cancel{m^2}}}{\underset{3}{\cancel{18km}}} = \frac{4km}{15}$$

You can also cancel binomial factors as long as the factors are identical. After you finish multiplying, use FOIL or the distributive property to simplify.

MULTIPLYING RATIONAL EXPRESSIONS WITH BINOMIALS

Example Multiply: $\frac{(x+3)(x-2)}{(x+1)} \cdot \frac{(x+5)}{(x-2)}$

Step 1
Cancel.

$$\frac{(x+3)\cancel{(x-2)}}{(x+1)} \cdot \frac{(x+5)}{\cancel{(x-2)}}$$

Step 2
Multiply straight across and simplify.

$$\frac{(x+3)\cancel{(x-2)}}{(x+1)} \cdot \frac{(x+5)}{\cancel{(x-2)}} = \frac{(x+3)(x+5)}{(x+1)} = \frac{x^2+8x+15}{x+1}$$

The numerators and denominators must be fully factored before you can cancel.

FACTORING RATIONAL EXPRESSIONS

Example Multiply: $\frac{x^2-12x+35}{x^2+2x} \cdot \frac{x^2-4}{x^2-6x-7}$

Step 1
Factor.

$$\frac{x^2-12x+35}{x^2+2x} \cdot \frac{x^2-4}{x^2-6x-7} = \frac{(x-7)(x-5)}{x(x+2)} \cdot \frac{(x-2)(x+2)}{(x-7)(x+1)}$$

Step 2
Cancel and then multiply. Simplify as needed. Leave the denominator factored.

$$\frac{\cancel{(x-7)}(x-5)}{x\cancel{(x+2)}} \cdot \frac{(x-2)\cancel{(x+2)}}{\cancel{(x-7)}(x+1)} = \frac{(x-5)(x-2)}{x(x+1)} = \frac{x^2-7x+10}{x(x+1)}$$

Multiply.

1. $\frac{3}{2v} \cdot \frac{5v^2}{2}$ $\qquad$ $\frac{9m}{2n} \cdot \frac{6}{7m^2}$ $\qquad$ $\frac{4y^2}{10xy} \cdot \frac{6x^2}{3y^2}$

2. $\frac{x+9}{2(x-3)} \cdot \frac{2(x-3)}{3}$ $\qquad$ $\frac{1}{n-1} \cdot \frac{(n+5)(n-6)}{n-6}$ $\qquad$ $\frac{1}{p-2} \cdot \frac{7p(p-2)}{9}$

3. $\frac{n+7}{n^2+2n-80} \cdot \frac{n^2+15n+50}{n+5}$ $\qquad$ $\frac{y-9}{9y+63} \cdot \frac{y^2+13y+42}{y-9}$ $\qquad$ $\frac{p^2+10p+21}{p+3} \cdot \frac{5p}{2p^3+14p^2}$

4. $\frac{v^2-9}{26v} \cdot \frac{2v-16}{v^2-5v-24}$ $\qquad$ $\frac{m-11}{2m^2-11m+14} \cdot \frac{m^2-13m+22}{m^2-22m+121}$ $\qquad$ $\frac{3x^2-14x-5}{x^2-19x+90} \cdot \frac{2x^2-13x-45}{3x^2-17x-6}$

5. $\frac{1}{n^2-15n+36} \cdot \frac{n^2-8n-48}{n+4}$ $\qquad$ $\frac{15x^2-60x}{x^2+21x-100} \cdot x^2+28x+75$ $\qquad$ $\frac{x^2+18x+65}{x^2+10x-39} \cdot \frac{x-3}{x+5}$

CORE CONNECTIONS: Rational Expressions and Domain

When you find the domain of a function, you look for all the possible values of *x*. The domain can't include any *x*-values that would result in a denominator of zero or a negative number inside a radical.

A rational expression is a fraction. To find the domain of a rational expression, look for values that would cause the denominator to equal zero. These values are called **excluded values** because they must be excluded from the domain.

Example What is the domain of the function $f(x) = \frac{3x^2}{x-9}$?

Set the denominator of the fraction equal to zero and solve for *x*.

$$x - 9 = 0$$
$$x = 9$$

The domain cannot include 9. The domain of $f(x) = \frac{3x^2}{x-9}$ is **all real numbers except for 9**, or $\boldsymbol{x \neq 9}$.

Find the domain of the following functions.

1. $f(x) = \frac{x-7}{x+3}$

2. $f(x) = \frac{4}{15x}$

3. $f(x) = \frac{6x+1}{2x+3}$

4. $f(x) = \frac{9}{(x-4)(x+5)}$

(***Hint:*** What happens when you multiply by zero? Set each factor equal to zero to find the excluded values.)

Answers start on page 182.

DIVIDE RATIONAL EXPRESSIONS

You know that division is the opposite of multiplication. When you divide fractions, you invert the fraction you are dividing by and then multiply.

Example Divide: $\frac{7}{12} \div \frac{5}{6}$

Step 1
Change to multiplication and invert the divisor.

$$\frac{7}{12} \div \frac{5}{6} = \frac{7}{12} \times \frac{6}{5}$$

Step 2
Cancel and multiply.

$$\frac{7}{\overset{}{\underset{2}{\cancel{12}}}} \times \frac{\overset{1}{\cancel{6}}}{5} = \frac{7}{10}$$

Dividing rational expressions works the same way. Change to multiplication and invert the second expression.

DIVIDING RATIONAL EXPRESSIONS WITH MONOMIALS

Example Divide: $\frac{m^3n}{20n} \div \frac{2mn}{15mn^3}$

Step 1
Change to multiplication and invert the divisor.

$$\frac{m^3n}{20n} \div \frac{2mn}{15mn^3} = \frac{m^3n}{20n} \cdot \frac{15mn^3}{2mn}$$

Step 2
Cancel.

$$\frac{m^3\cancel{n}}{\underset{4}{\cancel{20n}}} \cdot \frac{\overset{3n^2}{\cancel{15mn^3}}}{2\cancel{mn}}$$

Step 3
Multiply straight across.

$$\frac{m^3\cancel{n}}{\underset{4}{\cancel{20n}}} \cdot \frac{\overset{3n^2}{\cancel{15mn^3}}}{2\cancel{mn}} = \frac{3m^3n^2}{8}$$

Remember to factor each numerator and denominator before canceling. You can cancel identical factors only.

DIVIDING RATIONAL EXPRESSIONS WITH BINOMIALS

Example Divide: $\frac{x^2 - 8x}{x^2 + 8x + 16} \div \frac{x - 8}{x^2 - 2x - 24}$

Step 1
Change to multiplication and invert the divisor.

$$\frac{x^2 - 8x}{x^2 + 8x + 16} \cdot \frac{x^2 - 2x - 24}{x - 8}$$

Step 2
Factor.

$$\frac{x(x - 8)}{(x + 4)(x + 4)} \cdot \frac{(x + 4)(x - 6)}{x - 8}$$

Step 3
Cancel.

$$\frac{x\cancel{(x - 8)}}{(x + 4)\cancel{(x + 4)}} \cdot \frac{\cancel{(x + 4)}(x - 6)}{\cancel{x - 8}}$$

Step 4
Multiply across.

$$\frac{x\cancel{(x - 8)}}{(x + 4)\cancel{(x + 4)}} \cdot \frac{\cancel{(x + 4)}(x - 6)}{\cancel{x - 8}}$$

$$= \frac{x(x - 6)}{x + 4}$$

$$= \frac{x^2 - 6x}{x + 4}$$

Divide.

1. $\frac{5}{8} \div \frac{7x^2}{4x}$ $\qquad$ $\frac{6k^2}{8m} \div \frac{10k}{6m^2}$ $\qquad$ $\frac{14ab^2}{3} \div \frac{28b^2}{a^2}$

2. $\frac{1}{a+8} \div \frac{a+6}{2a^2(a+8)}$ $\qquad$ $\frac{9}{9(x-4)} \div \frac{10x^2}{(x+6)(x-4)}$ $\qquad$ $\frac{8m+12}{7m} \div \frac{8m+12}{m-1}$

3. $\frac{k^2+10k+9}{6k^3+6k^2} \div \frac{k+9}{9k}$ $\qquad$ $\frac{r^2-13r+36}{r-9} \div \frac{4r}{r+8}$ $\qquad$ $\frac{10n}{n^2+2n-80} \div \frac{10n}{7n+70}$

4. $\frac{y^2-6y-72}{y^2-24y+144} \div \frac{y^2+9y+18}{y}$ $\qquad$ $\frac{m^2-2m-3}{m^2-15m+50} \div \frac{4m^3-12m^2}{m^2-25}$ $\qquad$ $\frac{v^2+v-72}{v^2-17v+42} \div \frac{v^2+12v+27}{v^2-v-6}$

CORE CONNECTIONS: Complex Fractions

When you write a fraction, you are writing a division problem. You can use division to turn a fraction into a decimal. $\frac{3}{4}$ is the same as saying $3 \div 4$, and $3 \div 4 = 0.75$.

In algebra, you will sometimes see a fraction within a fraction. This is called a **complex fraction**. Use division to simplify complex fractions.

Example Simplify: $\frac{8}{\frac{1}{3}}$

Step 1
Turn the fraction into a division problem.

$$\frac{8}{\frac{1}{3}} = 8 \div \frac{1}{3}$$

Step 2
Divide.

$$\frac{8}{1} \div \frac{1}{3}$$
$$= \frac{8}{1} \bullet \frac{3}{1}$$
$$= 24$$

Simplify the following complex fractions.

1. $\frac{14}{\frac{2}{3}}$

2. $\frac{\frac{4}{5}}{\frac{6}{35}}$

3. $\frac{1}{1+\frac{1}{2}}$ (*Hint:* Add the numbers in the denominator first.)

4. $\frac{2+\frac{1}{3}}{3-\frac{1}{3}}$

Answers start on page 182.

SIMPLIFY RATIONAL EXPRESSIONS

When you simplify fractions, you look for the greatest common factor of the numerator and the denominator.

Use the same principle to simplify rational expressions. First, completely factor the numerator and the denominator. Then cancel any common factors. Rational expressions are fully simplified when you have canceled all common factors.

> **TIP**
>
> *For multiple-choice items, if your answer doesn't match any of the choices, you may still need to simplify your answer.*

SIMPLIFYING RATIONAL EXPRESSIONS

Example Simplify: $\frac{2x^2 - 18x + 36}{x^2 - 10x + 24}$

Step 1
Factor. Note that 2 is factored out of the numerator before factoring the trinomial.

$$\frac{2x^2 - 18x + 36}{x^2 - 10x + 24} = \frac{2(x^2 - 9x + 18)}{(x-4)(x-6)} = \frac{2(x-3)(x-6)}{(x-4)(x-6)}$$

Step 2
Cancel.

$$\frac{2(x-3)\cancel{(x-6)}}{(x-4)\cancel{(x-6)}} = \frac{2(x-3)}{(x-4)}$$

Step 3
Use the distributive property.

$$\frac{2(x-3)}{(x-4)} = \frac{2x-6}{x-4}$$

A rational expression can't have a denominator equal to zero. When a value results in zero in the denominator, the expression is **undefined** at that value. You must find the undefined values in a rational expression before you simplify.

You can find the undefined values of a rational expression by factoring. When you multiply by 0, the answer is always 0. Set each factor in the denominator equal to 0, and then solve for the variable. The solutions are the undefined values.

FINDING UNDEFINED VALUES

Example For which values is the expression $\frac{x^2 + x}{x^2 - 4x - 5}$ undefined?

Step 1
Factor.

$$\frac{x^2 + x}{x^2 - 4x - 5} = \frac{x(x+1)}{(x-5)(x+1)}$$

Step 2
Set each factor equal to 0 and solve for x.

$x - 5 = 0$ $\quad$ $x + 1 = 0$

$x = 5$ $\quad$ $x = -1$

The expression $\frac{x^2 + x}{x^2 - 4x - 5}$ is undefined when x is **5** or **–1**, so $x \neq 5$ and $x \neq -1$.

Evaluating a rational expression is easier if you simplify first. Perform the indicated operation. Then substitute the given value.

EVALUATING RATIONAL EXPRESSIONS

Example Evaluate $\frac{n^2+3n-40}{2n^2+7n} \div \frac{n^2+14n+48}{n^2}$ when $n = 4$.

Step 1
Change to multiplication and invert the divisor.

$$\frac{n^2+3n-40}{2n^2+7n} \cdot \frac{n^2}{n^2+14n+48}$$

Step 2
Factor.

$$\frac{(n+8)(n-5)}{n(2n+7)} \cdot \frac{n^2}{(n+8)(n+6)}$$

Step 3
Multiply.

$$\frac{\cancel{(n+8)}(n-5)}{\cancel{n}(2n+7)} \cdot \frac{\overset{n}{\cancel{n^2}}}{\cancel{(n+8)}(n+6)}$$

$$= \frac{n(n-5)}{(2n+7)(n+6)}$$

Step 4
Substitute and simplify.

$$\frac{n(n-5)}{(2n+7)(n+6)}$$

$$= \frac{4(4-5)}{(2 \cdot 4+7)(4+6)}$$

$$= \frac{-4}{150}$$

$$= -\frac{2}{75}$$

Simplify.

1. $\frac{x^2-x-2}{8x-16}$ $\qquad$ $\frac{n^2-11n+24}{n^2-8n+15}$ $\qquad$ $\frac{a^2+3a+2}{a^2-6a-7}$

2. $\frac{k^2+5k+4}{10k^2+40k}$ $\qquad$ $\frac{5x^3-40x^2}{2x^2-21x+40}$ $\qquad$ $\frac{r+3}{2r^2-4r-30}$

Find the values that make each expression undefined. Then simplify.

3. $\frac{x-5}{x^2-16x+55}$ $\qquad$ $\frac{y^2-10y+21}{y^2-7y}$ $\qquad$ $\frac{n^2+n-30}{n^2+4n-12}$

4. $\frac{n^2+5n-50}{n^2-25}$ $\qquad$ $\frac{3x^2+27x}{x^2+18x+81}$ $\qquad$ $\frac{m^2+6m-16}{2m^2+17m+8}$

Evaluate.

5. $\frac{u^2+1}{u^3}$ when $u = 3$

6. $\frac{2n-4}{2n^2+3n-14}$ when $n = -3$

7. $\frac{4m}{m+4} + \frac{m}{m-5}$ when $m = -4$

8. $\frac{a^2-3a-4}{a^2}$ when $a = 0$

9. $\frac{3x^2+12x+12}{x^2-7x-18}$ when $x = 1$

10. $\frac{y-1}{3y^2-2y-16} \cdot \frac{y^2-8y-20}{y^2-11y+10}$ when $y = 2$

Answers start on page 182.

Unit 5 Review

Solve as directed.

1. Which of the following correctly completes the factorization below?

$4x^2 + 27x + 18 = (4x + 3)(\_\_\_\_\_\_\_)$

A. $x + 6$

B. $x + 9$

C. $x + 15$

D. $4x + 6$

2. Evaluate $\frac{y^2 + 7y - 18}{2y^2 - 9y + 10}$ when $y = 3$.

A. 1

B. 5

C. 8

D. 12

3. What is the common denominator for the expressions $\frac{m+3}{m^2n}$ and $\frac{2-n}{mn^2}$?

A. mn

B. $m - n + 5$

C. m^2n^2

D. $(m + 3)(2 - n)$

4. Multiply: $\frac{v}{v^2 - 3v - 18} \cdot \frac{v^2 - 6v}{v + 2}$

A. $\frac{1}{v + 11}$

B. $\frac{v}{v + 3}$

C. $\frac{1}{(v + 2)(v + 3)}$

D. $\frac{v^2}{(v + 2)(v + 3)}$

5. For which of the following values is the expression $\frac{x^2 + 8x + 15}{x^2 - 6x - 40}$ undefined?

A. $x = -3$

B. $x = -5$

C. $x = 4$

D. $x = 10$

6. Which of the following pairs of terms are like terms?

A. ab and a^2b

B. $8b^2$ and $3b^2$

C. a^3 and a

D. a^2 and ab

Classify the following polynomial expressions. Select your answer.

7. $3a - 8$

- monomial
- binomial
- trinomial

8. $7u^2v^3$

- monomial
- binomial
- trinomial

9. $5r^2 + 3r - 9$

- monomial
- binomial
- trinomial

10. $2x^2 - 12xy$

- monomial
- binomial
- trinomial

Factor the following polynomial expressions.

11. $20x^2 - 45x$

12. $x^2 - 9x - 52$

Solve as directed.

13. Divide: $\frac{4x^2 - 12x}{x^2 - 49} \div \frac{9x - 27}{x^2 + 15x + 56}$

A. $\frac{x + 8}{9}$

B. $-\frac{4x + 8}{57}$

C. $\frac{4x^2 + 32x}{9x - 63}$

D. $\frac{36x}{x^2 + x - 56}$

14. Write the factor that completes the following factorization.

$72m^3n - 36m^2n - 48mn = \_\_\_\_\_\_(6m^2 - 3m - 4)$

15. If the area of the rectangle below is $2x^2 + x - 28$, which expression represents the length of the rectangle?

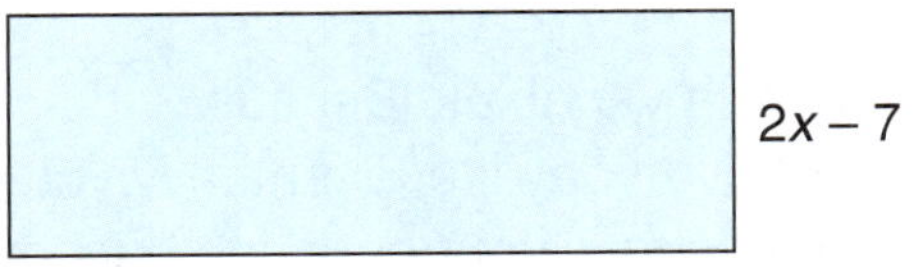

A. $2x^2 + 3x - 35$

B. $2x^2 - x - 21$

C. $2x + 4$

D. $x + 4$

16. Subtract: $\frac{y}{y + 3} - \frac{2}{y + 10}$

A. $\frac{y^2 + 8y - 6}{y^2 + 13y + 30}$

B. $\frac{10y - 6}{y^2 + 13y + 30}$

C. $\frac{10y - 6}{2y + 13}$

D. $\frac{y - 2}{2y + 13}$

17. Factor: $m^2 + 2m + 1$

18. Evaluate $\frac{3r^2 - 10r}{3r^2 - 4r - 20}$ when $r = 0$.

19. Add: $\frac{u + v}{u^2} + \frac{u}{v}$

A. $\frac{v}{u^2v}$

B. $\frac{2u + v}{u^2 + v}$

C. $\frac{2u + v}{u^2v}$

D. $\frac{u^3 + uv + v^2}{u^2v}$

20. For which values is the expression $\frac{6x^2 - 42x}{x^2 - 12x + 35}$ undefined?

A. $x = 5$

B. $x = 0$, $x = 7$

C. $x = 5$, $x = 7$

D. $x = 0$, $x = 5$, $x = 7$

21. Simplify: $(3m^2 - mn + 5) - (m^2 + 12)$

22. **Explain** Zoe evaluates the expression $\frac{x^2 - 5x - 24}{x^2 + 18x + 45}$ when $x = 3$ and gets the answer $-\frac{5}{18}$. Sean says that the expression is undefined when $x = 3$. Who is correct? Explain your thinking.

Answers start on page 182.

UNIT

6

QUADRATIC EQUATIONS

SKILLS

- Taking the square root of both sides of an equation
- Analyzing the graph of a quadratic
- Solving quadratic equations by factoring
- Completing the square
- Solving word problems with quadratic equations

TOOLS AND STRATEGIES

- Graphing a quadratic equation
- Using the quadratic formula

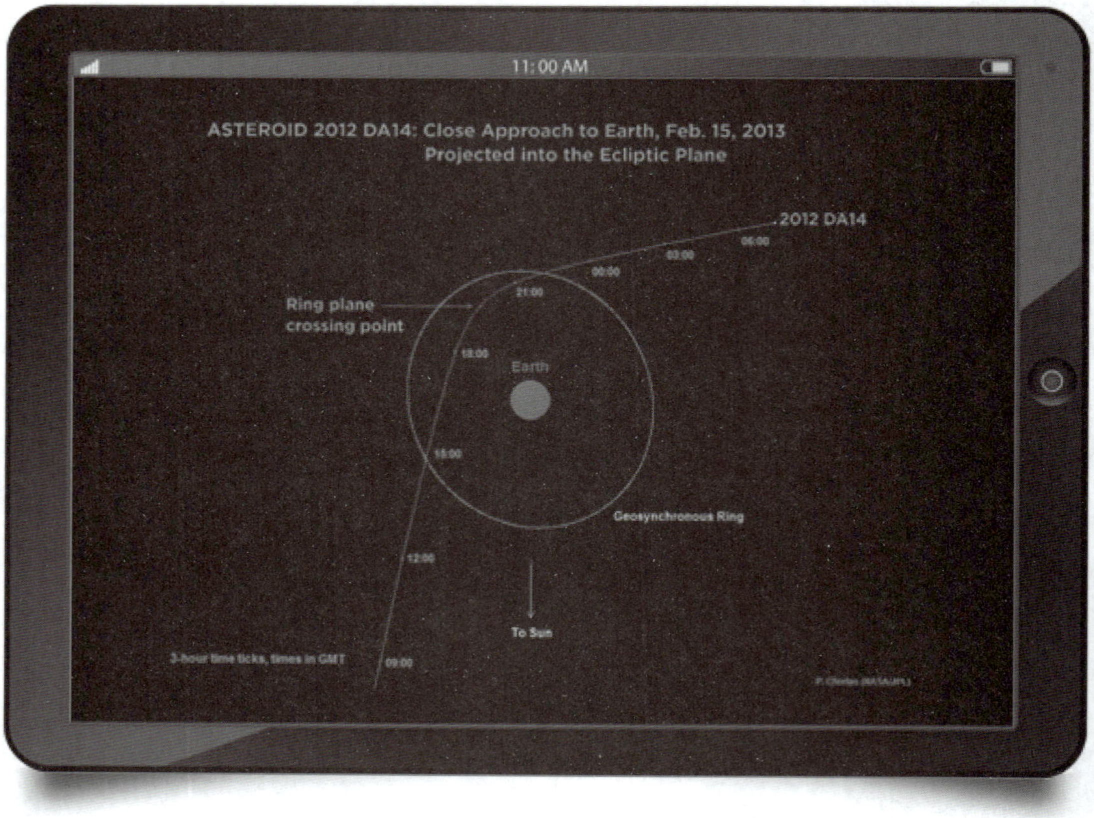

MANY REAL-WORLD SITUATIONS are modeled using multiplication. For example, you can figure out how far you can drive in a number of hours by multiplying the hours by your average speed. You can find the area of any rectangle by multiplying length and width. You can find the total cost by multiplying the number of items by the price per item.

In algebra, you can represent quantities with variables. Because multiplication is such an important part of everyday life, eventually you will need to multiply one x by another x. Immediately, your equation becomes quadratic.

Without special methods for solving quadratic equations, you would have no choice but to use trial and error. Luckily, people have developed four different ways of solving quadratic equations. You might be surprised to learn that these methods are thousands of years old. In this unit, you will learn how to use these methods.

USING QUADRATIC EQUATIONS

You use quadratic equations when you:

- Solve problems where area is increasing or decreasing
- Calculate how far a projectile will travel in a certain amount of time
- Apply a formula where a variable is raised to the second power
- Solve business equations about profit and loss
- Use functions to model things that happen in the world

Read the following items. Describe some of your experiences in writing or discuss them with a partner.

1. While playing a sport, have you ever had to catch a ball thrown or hit high into the air? How did you know where the ball would come down?

2. Have you ever wondered how scientists can predict what will happen when they send a rocket into space? What are some of the factors that you imagine they must consider to make accurate predictions?

3. How do you decide how much room to leave between your car and another car while driving?

4. Have you ever waited to buy something, hoping it would go on sale? Businesses use quadratic equations to decide what price to put on a product. If there is high demand for a product, the business can charge more. If people don't want it, the price goes down. How could you use this idea to decide when to buy something you really want?

MATH TALK

Imagine this situation: You stake out a vegetable garden that measures 20 meters by 20 meters. The area of the garden is 400 square meters. Later, you decide you want the garden to have 100 more square meters. By how much would you have to lengthen each side to achieve your goal?

Discuss with a partner how you could figure out what the length of each side of the square garden should be.

More than 2,000 years ago, situations like this one led mathematicians in Egypt to develop a method called *completing the square*. You will learn this method and others in this unit.

QUADRATIC EQUATION BASICS

A **quadratic equation** has a variable that is squared, or raised to a power of 2. No variable can have an exponent greater than 2.

A **power** shows repeated multiplication of the same number. The expression 5^2 is read "5 to the second power" or "5 squared." The number 5 is the **base** and 2 is the **exponent**. The exponent tells how many times the base appears in the multiplication problem.

What is the value of x^2 when x equals 5? $x^2 = x \cdot x$ $5^2 = 5 \cdot 5 = \mathbf{25}$

What is the value of x^2 when x equals –5? $(-5)^2 = -5 \cdot -5 = \mathbf{25}$

As you can see, both 5^2 and $(-5)^2$ have the same value.

A **square root** does the opposite of squaring. The $\sqrt{}$ symbol indicates square root. The expression $\sqrt{x^2}$ is equal to x. The expression $\sqrt{25}$ is equal to both 5 and –5.

FINDING SQUARE ROOTS TO SOLVE A QUADRATIC EQUATION

Example 1 Solve for x. $x^2 = 36$

Find the square root of both sides.

$\sqrt{x^2} = \sqrt{36}$

$x = 6$ and -6

There are two solutions. You can write the solution set in braces: **{6, –6}**

Example 2 Solve for y. $y^2 - 5 = 44$

Step 1
Add 5 to both sides.

$y^2 - 5 = 44$

$y^2 = 49$

Step 2
Find the square root of both sides.

$\sqrt{y^2} = \sqrt{49}$

$y = 7$ and -7

The solution set is **{7, –7}**.

The solutions for a quadratic equation will not always be integers. The solution set may contain fractions. You may also need to leave the radical symbol $\sqrt{}$ in the answer and simplify the expression. Study the following examples to see how to work with answers that are not integers.

Example with fractions:

$$9x^2 = 4$$
$$x^2 = \frac{4}{9}$$
$$\sqrt{x^2} = \sqrt{\frac{4}{9}}$$
$$x = \frac{2}{3} \text{ and } -\frac{2}{3}$$

Example with roots in the answer:

$$n^2 = 12$$
$$\sqrt{n^2} = \sqrt{12}$$
$$\sqrt{n^2} = \sqrt{4 \cdot 3}$$
$$n = 2\sqrt{3} \text{ and } -2\sqrt{3}$$

Always simplify answers as much as possible. To review how to simplify radicals, see *Math Sense 2: Focus on Problem Solving,* pp. 54–55.

Solve for the variable. Write the solution set in braces.

1. $k^2 = 81$ $\quad$ $n^2 = 0$ $\quad$ $m^2 = 100$

2. $x^2 + 10 = 74$ $\quad$ $p^2 - 10 = 8$ $\quad$ $8y^2 = 32$

3. $a^2 - 2 = 8$ $\quad$ $81x^2 + 5 = 14$ $\quad$ $36b^2 + 2 = 83$

In a real-world situation, you may need to use a calculator to evaluate a root. Use these keystrokes to find the value of $\sqrt{13}$ on the GED® online calculator.

[2nd] $\sqrt{\ }$ 13 [enter] [◄ ►]

$\sqrt{13}$
$\sqrt{13}$ ◄►

$\sqrt{13}$
3.605551275

$\sqrt{13}$ is equal to about **3.6**.

Solve using a calculator. Round your answer to the tenths place.

4. $x^2 = 23$ $\quad$ $n^2 - 5 = 49$ $\quad$ $3m^2 - 1 = 284$

5. $1 - 3b^2 = -29$ $\quad$ $2k^2 - 3 = 27$ $\quad$ $7y^2 + 4 = 82$

6. $3x^2 + 1 = 10$ $\quad$ $5 - 4n^2 = -23$ $\quad$ $2m^2 + 3 = 121$

Answers start on page 183.

SOLVE QUADRATIC EQUATIONS USING GRAPHS

Tools

The graph of a quadratic equation forms a **parabola**, a symmetrical curved line. A parabola can open upward or downward.

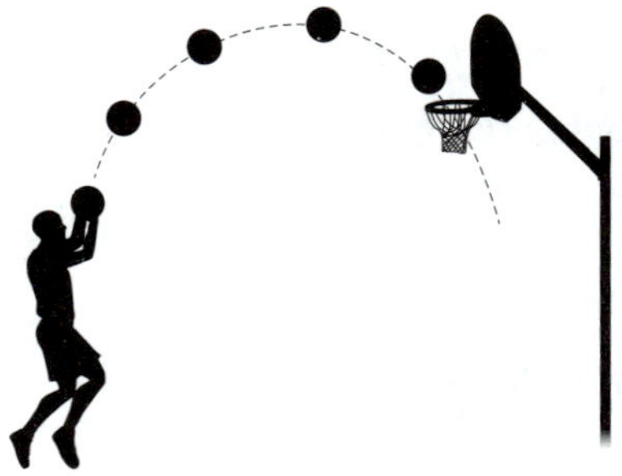

When you throw a ball into the air, the path of the ball is a parabola.

To graph a quadratic equation, rewrite the equation so that one side equals 0. Then replace 0 with *y*.

Solve $x^2 = 4$.	You already know the equation has two answers: 2 and –2.
Subtract 4 from both sides.	$x^2 - 4 = 0$
Replace 0 with *y*.	$x^2 - 4 = y$, which can be written $y = x^2 - 4$

Now you are ready to graph the equation.

GRAPHING A QUADRATIC EQUATION

Example Graph: $y = x^2 - 4$

Step 1
Make a table of values. Choose values for *x*. Substitute them into the equation and solve for *y*.

x	*y*
–3	5
–2	0
–1	–3
0	–4
1	–3
2	0
3	5

Step 2
Plot the points from the table.

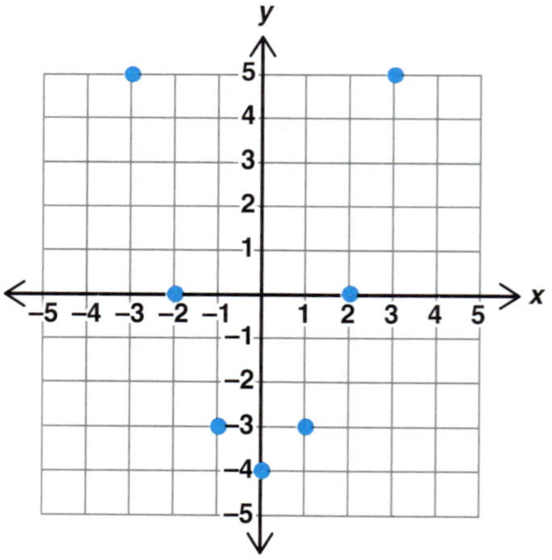

Step 3
Sketch the parabola.

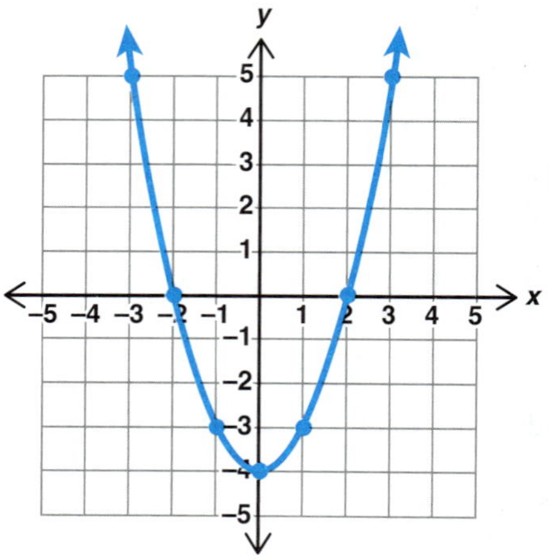

The graph tells important information about the equation.

The lowest or highest point on a parabola is its **vertex**.

Here, the vertex shows the lower limit of the range. The range of the equation is all real numbers greater than or equal to –4.

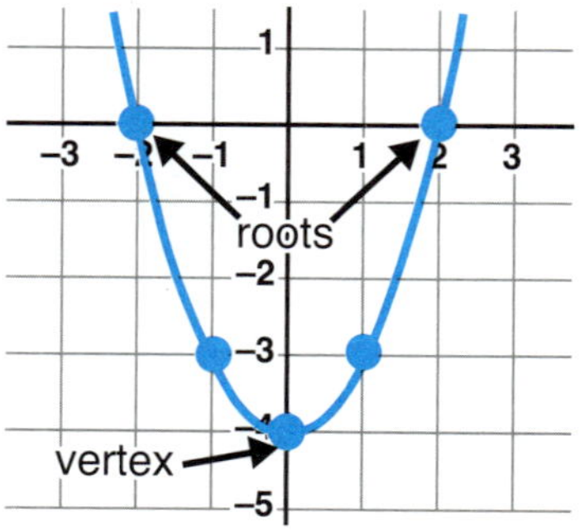

The **roots** are found where the graph crosses the *x*-axis. The roots are the solutions to the equation.

The solutions are –2 and 2.

You can also tell which way a parabola will open by looking at the term containing the squared variable.

- If the term is positive, the parabola opens upward.
- If the term is negative, the parabola opens downward.

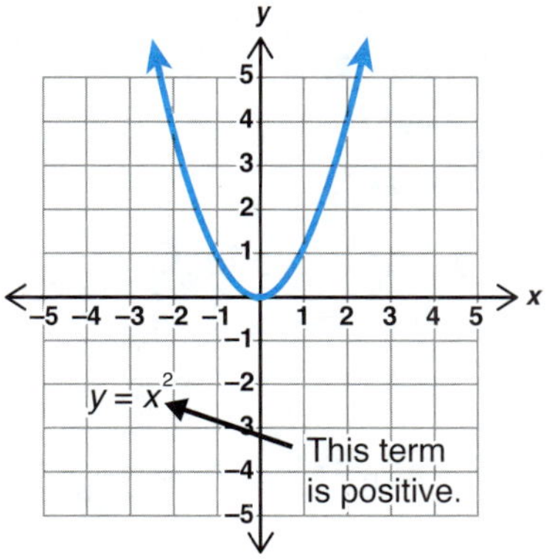

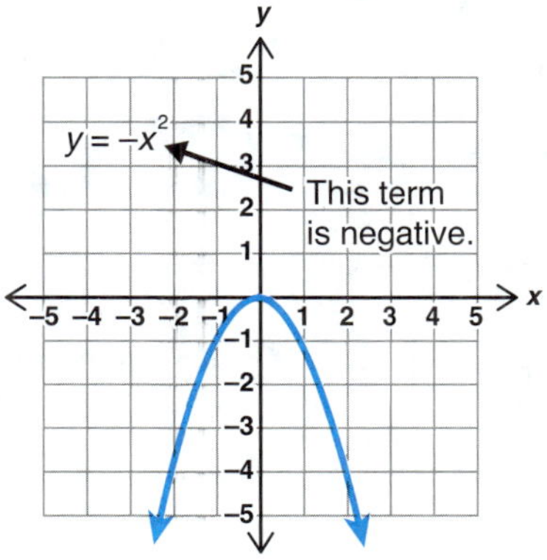

Solve as directed.

1. Graph the equation $y = -x^2 - 2x + 3$.

a. Create a table of values. Plot a minimum of 5 points.

b. What are the coordinates of the vertex?

c. What are the roots of the equation?

d. Find the domain and range of the equation.

x	*y*

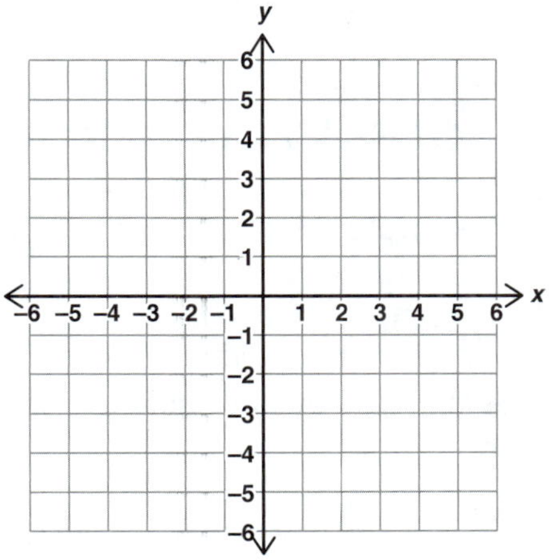

Match the equation to its graph. The vertex may not be visible in every graph.

2. $y = -x^2 + x + 6$

3. $y = -x^2 + 1$

4. $y = 2x^2 - 7x + 4$

5. $y = x^2 + 5x + 4$

A.

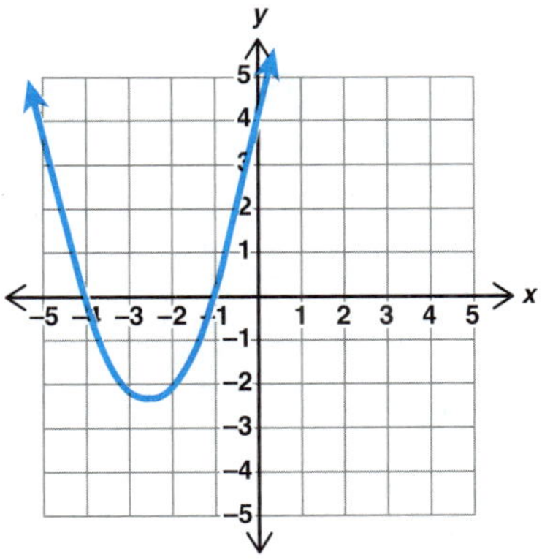

C.

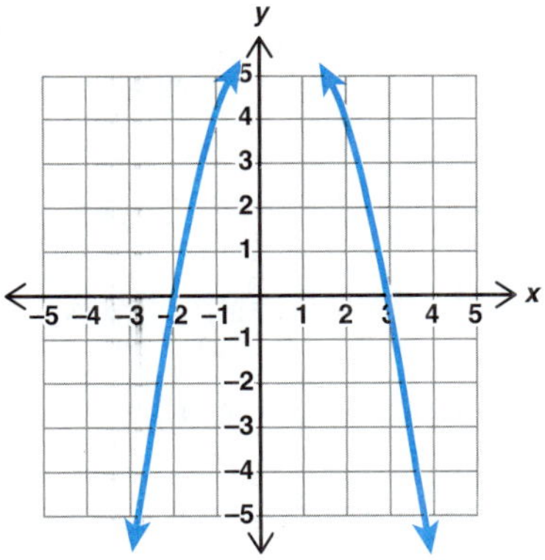

B.

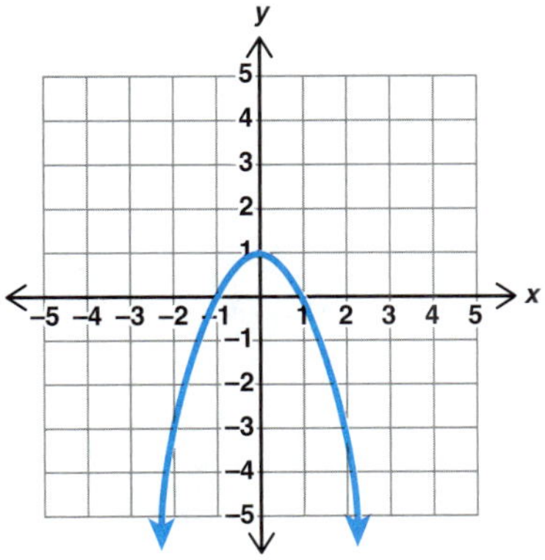

D.

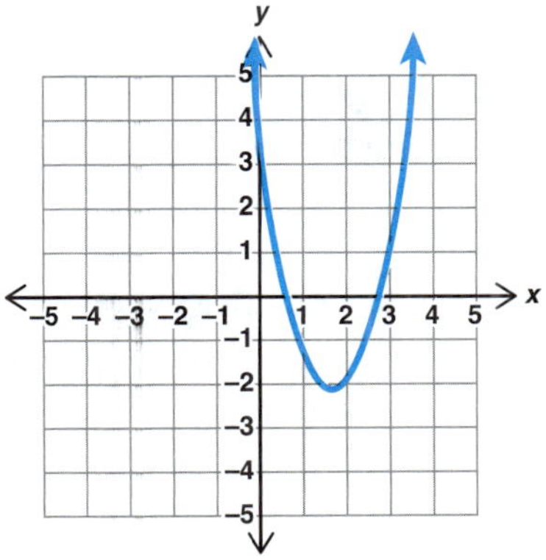

Answers start on page 183.

SOLVE QUADRATIC EQUATIONS BY FACTORING

When you factor an expression, you rewrite it as a product.

$ab + ac$ ⟵ This expression is the sum of two terms.

$a(b + c)$ ⟵ The factored expression is the product of a and the quantity $(b + c)$.

Factoring makes it possible to solve quadratic equations because of the **zero property of multiplication**. It states that when the product of two numbers is 0, at least one of the numbers must be 0. For example, if $ab = 0$, then either $a = 0$, $b = 0$, or both equal 0.

SOLVING BY FACTORING

Example Solve for x. $x^2 - 2x = 15$

Step 1
Set the quadratic expression equal to 0 by subtracting 15 from both sides.

$$x^2 - 2x = 15$$
$$x^2 - 2x - 15 = 0$$

Step 2
Factor the left side of the equation.

$$x^2 - 2x - 15 = 0$$
$$(x + 3)(x - 5) = 0$$

Step 3
The two factors are $(x + 3)$ and $(x - 5)$. The equation is true if either factor equals 0. Set the factors equal to 0, and solve for x.

$$x + 3 = 0 \qquad x - 5 = 0$$
$$x = -3 \qquad x = 5$$

The quadratic equation has two solutions: **{–3, 5}**. Either will make the original equation true.

$$x^2 - 2x = 15$$
$$(-3)^2 - 2(-3) = 15$$
$$9 + 6 = 15$$
$$15 = 15 \quad \text{True}$$

$$x^2 - 2x = 15$$
$$5^2 - 2(5) = 15$$
$$25 - 10 = 15$$
$$15 = 15 \quad \text{True}$$

Some equations have only one solution.

Example Solve for m. $m^2 - 8m + 16 = 0$

The equation is already set equal to 0.

Factor the left side:

$$m^2 - 8m + 16 = 0$$
$$(m - 4)(m - 4) = 0$$

Both factors are the same, so there is only one solution.

$$m - 4 = 0$$
$$m = 4$$

The solution set is **{4}.**

Check:

$$m^2 - 8m + 16 = 0$$
$$4^2 - 8(4) + 16 = 0$$
$$16 - 32 + 16 = 0$$
$$0 = 0 \quad \text{True}$$

Some quadratic equations cannot be solved by factoring. In the next two lessons, you will learn special techniques for solving more difficult quadratic equations. For now, continue to practice factoring.

Factor and solve for the variable. Write the solution set in braces.

1. $k^2 - 16 = 0$ $\quad$ $n^2 - 8n = 0$ $\quad$ $m^2 + 2m - 24 = 0$

2. $x^2 - 15x + 56 = 0$ $\quad$ $n^2 - n - 20 = 0$ $\quad$ $a^2 + 3a - 4 = 0$

3. $w^2 - 24 = 5w$ $\quad$ $k^2 - 12k = -35$ $\quad$ $b^2 = -28 + 11b$

   (***Hint:*** Set the equation equal to 0.)

4. $3x^2 + 23x - 8 = 0$ $\quad$ $2n^2 - 13n + 6 = 0$ $\quad$ $14m^2 + 9m + 1 = 0$

   (***Hint:*** Factor the first term as $7m$ and $2m$.)

5. $7n^2 - 34n = -24$ $\quad$ $5a^2 - 4a - 8 = -8$ $\quad$ $2c^2 - 7c + 1 = -2$

6. Which of the following equations is graphed on the coordinate plane?
   A. $y = x^2 + x - 6$
   B. $y = x^2 - 4x + 4$
   C. $y = x^2 - 2x$
   D. $y = x^2 - x$

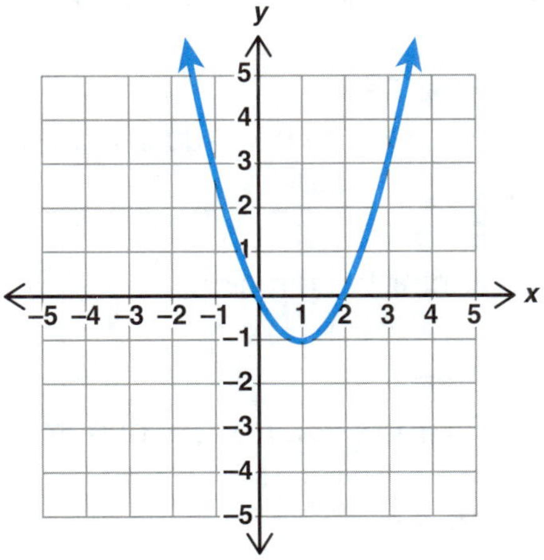

7. **Explain** How could factoring help you match the correct equation to the graph?

Answers start on page 183.

COMPLETING THE SQUARE

Some trinomials are perfect squares. When they are factored, the two binomial factors are identical. These trinomials are called **trinomial squares**.

$x^2 + 10x + 25$
$= (x + 5)^2$

Completing the square is a technique for solving quadratic equations that uses trinomial squares. Study the example to see how to make a trinomial square.

Example Fill in the blank to make a perfect square. $x^2 + 8x +$ _____

Step 1
Take the number from the middle term: 8

Step 2
Divide by 2.

$8 \div 2 = 4$

Step 3
Square the result.

$4^2 = 16$

Step 4
Fill in the blank.

$x^2 + 8x + 16$

Now you are ready to solve an equation by completing the square.

SOLVE BY COMPLETING THE SQUARE

Example Solve for *x*. $x^2 - 12x + 20 = 7$

Step 1
Make a space to complete the square.
Subtract 20 from both sides of the equation.

$x^2 - 12x + 20 = 7$
$x^2 - 12x \_\_\_\_\_ = -13$

Step 2
What number completes the perfect square?
Divide the number of the middle term by 2. Square the result.

$-12 \div 2 = -6$
$(-6)^2 = \mathbf{36}$

Step 3
Fill in the blank. Add 36 to *both sides*.

$x^2 - 12x \_\_\_\_\_ = -13$
$x^2 - 12x + 36 = 23$

Step 4
Factor the left side.

$(x - 6)^2 = 23$

Step 5
Take the square root of both sides.

$\sqrt{(x - 6)^2} = \sqrt{23}$
$x - 6 = \pm\sqrt{23}$

Step 6
Isolate the variable. Add 6 to both sides.

$x = 6 \pm \sqrt{23}$

TIP
The symbol ± means "plus or minus."

There are two solutions: $\mathbf{6 + \sqrt{23}}$ and $\mathbf{6 - \sqrt{23}}$.

You can use your calculator to find the value of the solutions. Since the solutions represent irrational numbers, you will need to round to a convenient decimal place.

$6 + \sqrt{23} \approx 10.7958 \qquad 6 - \sqrt{23} \approx 1.2042$

Finish the steps to complete the square.

1. Solve $n^2 - 14n = -25$.

Complete the trinomial square. Add the same number to both sides. $n^2 - 14n + \_\_\_\_\_ = -25 + \_\_\_\_\_$

Factor the left side. $(n \_\_\_\_\_)^2 = \_\_\_\_\_$

Find the square root of both sides. $\sqrt{(n \_\_\_\_\_)^2} = \sqrt{\_\_\_\_\_}$

$n \_\_\_\_\_ = \pm\sqrt{\_\_\_\_\_}$

Write the two solutions in braces. $\{\_\_\_\_\_\_\_\_\_\_, \_\_\_\_\_\_\_\_\_\_\}$

All the problems below can be solved by completing the square. Some can be solved by factoring. Choose a method for each problem, and solve for the variable.

2. $x^2 + 14x + 37 = 4$

3. $n^2 + 10n - 60 = 6$

4. $a^2 - 6a - 28 = 0$

5. $x^2 + 10x - 69 = 0$

6. $m^2 - 14m - 48 = 3$

7. $p^2 + 8p - 37 = 10$

8. $m^2 + 18m + 59 = 0$

9. $n^2 - 18n + 85 = 5$

10. Solve for b. Write the solution set using decimals. Round your answers to the nearest thousandth.

$b^2 - 2b + 3 = 10$

Answers start on page 184.

USE THE QUADRATIC FORMULA

Tools

You've already learned three ways to solve a quadratic equation: graphing, factoring, and completing the square. A fourth way involves a formula. The quadratic formula looks complicated, but with practice, you will find it easy to use. The best thing about the quadratic formula is that it works in every situation.

Quadratic Formula When $ax^2 + bx + c = 0$, $x = \dfrac{-b \pm \sqrt{b^2 - 4ac}}{2a}$

SOLVE EQUATIONS USING THE QUADRATIC FORMULA

Example Solve for *x*. $3x^2 + 11x - 4 = 0$

Step 1

The equation is set equal to 0. Find the values for *a*, *b*, and *c*.

$3x^2 + 11x - 4 = 0$ $a = 3$ $b = 11$ $c = -4$

Step 2

Substitute the values into the quadratic formula and simplify.

$$x = \frac{-b \pm \sqrt{b^2 - 4ac}}{2a} \qquad \frac{-11 \pm \sqrt{11^2 - 4 \cdot 3 \cdot -4}}{2 \cdot 3} = \frac{-11 \pm \sqrt{121 + 48}}{6} = \frac{-11 \pm \sqrt{169}}{6}$$

Step 3

Use your calculator to evaluate the root. $\dfrac{-11 \pm \sqrt{169}}{6} = \dfrac{-11 \pm 13}{6}$

Step 4

Simplify the two roots. Use addition for one and subtraction for the other.

$$\frac{-11 + 13}{6} = \frac{2}{6} = \frac{1}{3} \qquad \frac{-11 - 13}{6} = \frac{-24}{6} = -4$$

The solution set is $\{\frac{1}{3}, -4\}$.

If the roots are irrational numbers, you will be left with a root symbol in your answer. Follow the steps to see what happens when you solve $m^2 - 4m - 6 = 0$.

$$x = \frac{-b \pm \sqrt{b^2 - 4ac}}{2a} \qquad \frac{-(-4) \pm \sqrt{-4^2 - 4 \cdot 1 \cdot -6}}{2 \cdot 1} = \frac{4 \pm \sqrt{16 + 24}}{2} = \frac{4 \pm \sqrt{40}}{2}$$

Simplify the root. $\dfrac{4 \pm \sqrt{40}}{2} = \dfrac{4 \pm 2\sqrt{10}}{2}$ You can divide each term by 2. $\dfrac{4 \pm 2\sqrt{10}}{2} = \dfrac{4}{2} \pm \dfrac{2\sqrt{10}}{2} = 2 \pm \sqrt{10}$

The solution set is $\{2 + \sqrt{10}, 2 - \sqrt{10}\}$.

You may be wondering whether you could use a calculator do the work. You can, but the keystrokes can be complicated. Good test takers do the algebra steps by hand and use a calculator to simplify the roots.

Solve using the quadratic formula. Remember to set the equation equal to 0 before you use the formula.

1. $x^2 + 4x - 21 = 0$

2. $2x^2 + 5x - 25 = 0$

3. $x^2 + 3x - 2 = 8$
(***Hint:*** Subtract 8 from both sides to set the equation equal to 0.)

4. $x^2 - 2 = 0$
(***Hint:*** $b = 0$ because there is no middle term.)

5. $3x^2 - 8x + 4 = 0$

6. $x^2 - 4x = 1$

7. $4x^2 + 2x - 3 = 0$

8. $x^2 - 18 = 0$

9. $3x^2 - 10x - 2 = 2$

10. $9x^2 - 8x - 5 = 0$

Solve as directed.

11. Use your calculator to find the roots of the equation in problem 10. Round the answers to the nearest tenth.

12. **Challenge** Without graphing, describe what the graph of the equation in problem 1 would look like.

Answers start on page 184.

SOLVE WORD PROBLEMS WITH QUADRATIC EQUATIONS

On the GED® Math Test, you may be asked to solve word problems that require quadratic expressions and equations. Sometimes, you will be given an equation to help you solve the problem.

Example 1 Jonas stands on the roof of a building, 50 meters above the ground. He throws a tennis ball into the air with an initial velocity of 15 meters per second. How many seconds will it take for the ball to hit the ground?

Use the formula $-5t^2 + vt + c = 0$, where t = time in seconds, v = initial velocity, and c equals the starting position of the ball.

Step 1
Substitute the given values into the formula.

$$-5t^2 + vt + c = 0$$
$$-5t^2 + 15t + 50 = 0$$

Step 2
Solve the equation for t, the time in seconds.

$$\frac{-b \pm \sqrt{b^2 - 4ac}}{2a} = \frac{-15 \pm \sqrt{15^2 - 4 \cdot -5 \cdot 50}}{2 \cdot -5}$$
$$= \frac{-15 \pm \sqrt{1225}}{-10} = \frac{-15 \pm 35}{-10}$$

Find the roots, adding in one and subtracting in the other.

$$\frac{-15 + 35}{-10} = -2 \quad \text{and} \quad \frac{-15 - 35}{-10} = 5$$

Step 3
Interpret the results. The ball can't be in the air for –2 seconds. In this real-world situation, a negative answer doesn't make sense. This is an **extraneous solution**. The other solution makes sense.

The ball will hit the ground in **5 seconds**.

In other problems, you will be expected to write an equation using the given information.

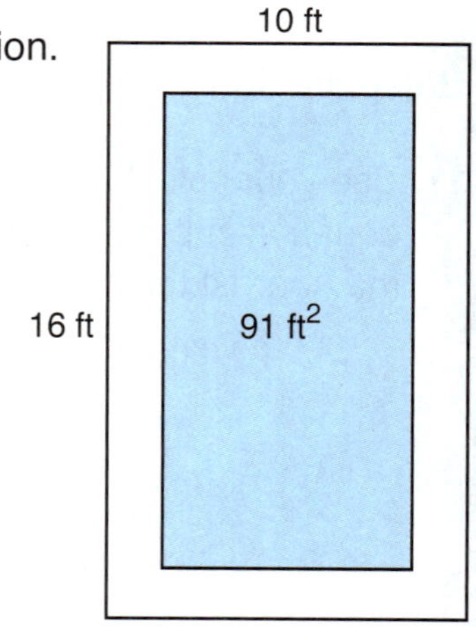

Example 2 Maria wanted a rose garden that was 16 feet long and 10 feet wide. However, she had to reduce the length and the width by the same number of feet. If the area of the new garden is 91 square feet, what are the dimensions of the new garden?

Step 1
Write an equation. The length and width are reduced by the same number, x.

$$(10 - x)(16 - x) = 91$$

Step 2
Use FOIL to write the quadratic equation. Solve for x. Factoring is used for this example.

$$160 - 10x - 16x + x^2 = 91$$
$$x^2 - 26x + 160 = 91$$
$$x^2 - 26x + 69 = 0$$
$$(x - 23)(x - 3) = 0$$

The roots are 23 and 3, but 23 doesn't makes sense in this situation. Maria reduced the length and width by 3 feet. The new dimensions are **7 feet** and **13 feet**.

In real life, a quadratic equation often has one answer that doesn't make sense in the situation. Good test takers always look back at the original problem to make sure the answer makes sense. Common sense is an important part of test taking.

Solve.

1. The sum of the square of a positive number and the number itself is 210. What is the number?

2. The amount of time it takes in seconds for a dropped object to hit the ground can be found using the equation $-16t^2 + c = 0$, where t is time in seconds and c is the initial height of the object in feet.

   For a science project, an egg is dropped from a window that is 90 feet from the ground. To the nearest hundredth, how many seconds will it take for the egg to hit the ground?

3. City planners decide to extend a square playground on the east and south sides by an equal distance. If the new playground will have an area of 784 square meters, what is the width of the extension?

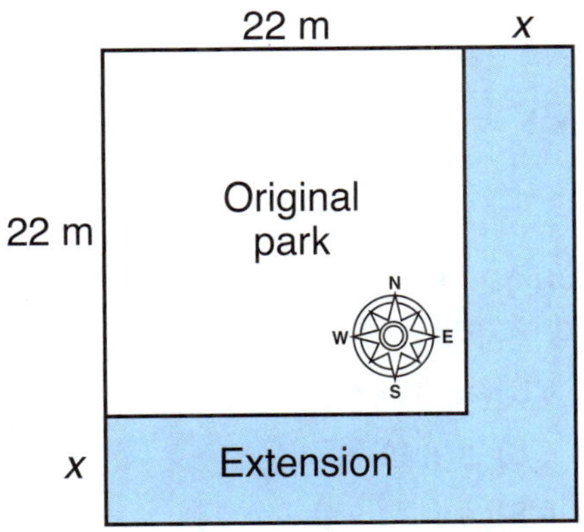

4. The area of a rectangular parking lot is 10,400 square meters. The length of the lot is 50 meters more than the width. What are the dimensions of the lot?

5. When the sides of a square are each increased by 9 inches, the area of the larger square is 369 square inches greater than the area of the original square. How many inches long is each side of the new square?

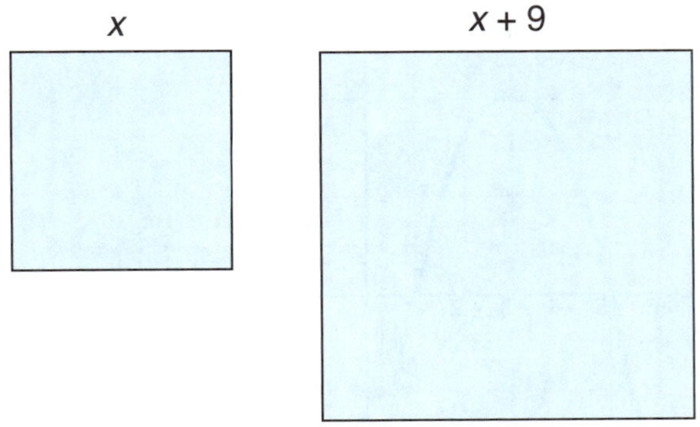

6. The distance (s) that an object travels can be found when the initial speed (v), the time elapsed (t), and the rate of constant acceleration (a) are known. Use this equation:

$$s = vt + \frac{1}{2}at^2$$

   If a scooter has an initial speed of 10 meters per second and a constant acceleration of 2 meters per second, how many seconds will it take the scooter to travel 600 meters?

Answers start on page 184.

Unit 6 Review

Solve as directed.

1. An artist wants to create a square base for a sculpture. She wants the base to have an area of 70 square inches. In inches, what is the approximate length of each side of the base?

   A. 5.9

   B. 8.4

   C. 8.75

   D. 17.5

Use the graph for problems 2–4.

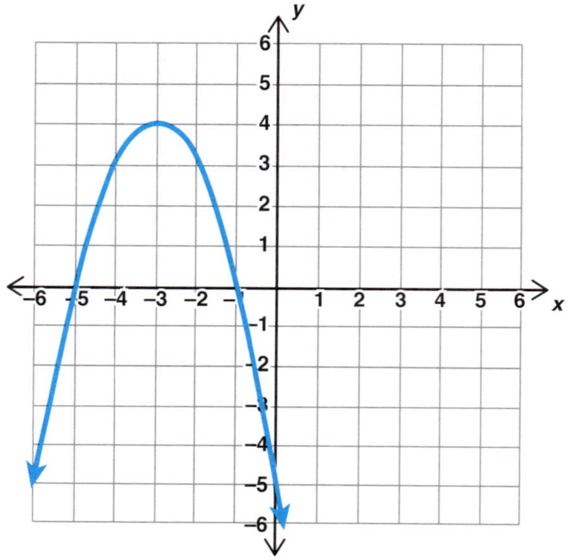

2. What is the vertex of the graph?

3. Select the roots of the graph.

(–6, –5)	(–5, 0)
(–4, 3)	(–3, 0)
(–1, 0)	(0, –5)

4. Which of the following points does not lie on the graph?

   A. (–5, 1)

   B. (–4, 3)

   C. (–2, 3)

   D. (0, –5)

5. Which equation matches the graph below?

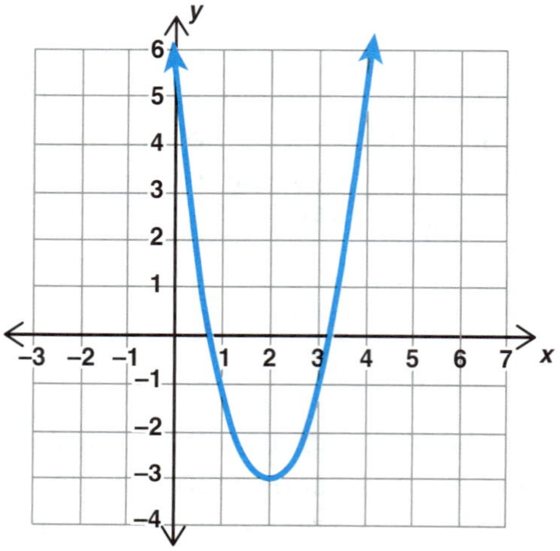

   A. $y = x^2 + 5$

   B. $y = x^2 - 4x + 1$

   C. $y = x^2 - 4x + 3$

   D. $y = 2x^2 - 8x + 5$

6. The solution set $\{-1, \frac{8}{3}\}$ is the solution to which factored equation?

   A. $(8a - 3)(a + 1) = 0$

   B. $(3a - 8)(a + 1) = 0$

   C. $(8a + 3)(a - 1) = 0$

   D. $(3a - 8)(a - 1) = 0$

7. Write the solution set to the equation below.

   $x^2 - 10 = 3x$

8. Gordon is solving the equation below by completing the square.

$$k^2 - 16k - 40 = 40$$

First, he adds 40 to both sides of the equation.

$$k^2 - 16k = 80$$

Next, he needs to create a trinomial square on the left side. What number does he add to both sides of the equation?

$$k^2 - 16k + \_\_\_\_ = 80 + \_\_\_\_$$

A. 32

B. 40

C. 64

D. 160

9. A softball player throws a ball into the air at an initial velocity of 44 feet per second. The player's hand is 5 feet from the ground at the time the ball is released.

Use the formula $-16t^2 + vt + c = 0$.

t = time in seconds the ball is in the air
v = initial velocity in feet per second
c = the initial distance from the ground
h = the distance above the ground in feet

How many seconds will the ball stay in the air? Round your answer to the nearest hundredth second.

A. 1.09

B. 1.59

C. 1.82

D. 2.86

10. The equation $2x^2 + 7x = 39$ has two solutions. Write the negative solution in the box below.

You may write your answer as a fraction or decimal.

[ ]

11. A company makes metal frames for 8- by 10-inch photographs. The area of each metal frame is 63 square inches. How wide is the frame in inches? (***Hint:*** If the area of the frame is 63 in^2, then the area of the frame and picture is 63 + 80 in^2.)

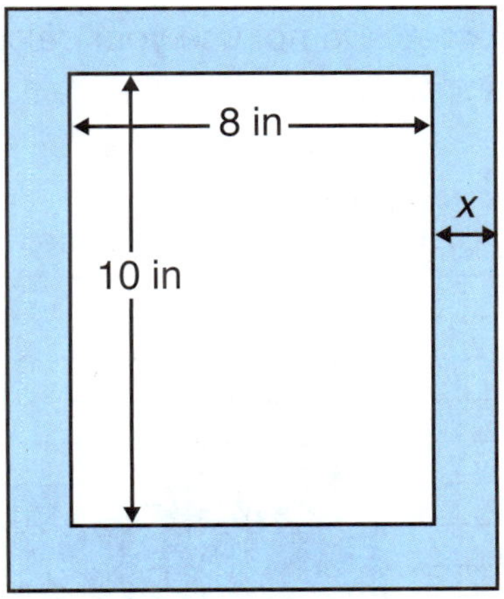

A. $\frac{3}{4}$

B. 1

C. $1\frac{1}{2}$

D. $1\frac{3}{4}$

12. The lawn in front of a courthouse is in the shape of a right triangle. One leg is 17 feet longer than the other. The hypotenuse of the triangular lawn is 53 feet.

What is the length of the shortest side of the lawn in feet? (***Hint:*** Use the Pythagorean theorem.)

[ ] feet

Answers start on page 185.

Simulated GED® Test

This 25-item simulated test will help you practice the skills you have learned in this book.

Solve as directed. Do not use your calculator for questions 1 and 2.

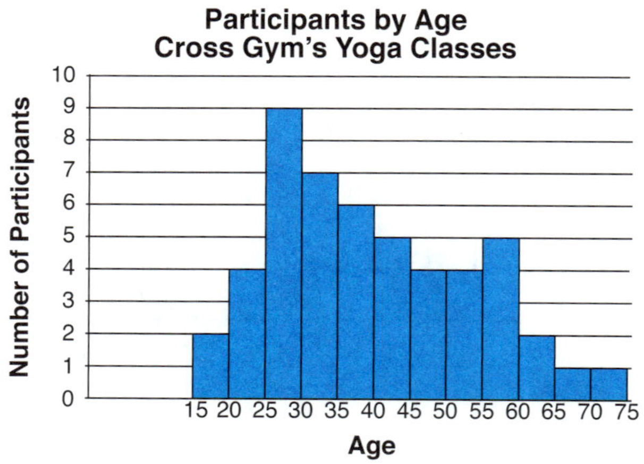

1. Of the 50 people who signed up for Cross Gym's Yoga Classes, how many are under 40 years of age?

A. 6

B. 10

C. 28

D. 33

2. Which correctly factors the following expression?

$$3x^2 + x - 24$$

A. $(3x + 8)(x - 3)$

B. $(3x - 8)(x + 3)$

C. $(3x + 4)(x - 6)$

D. $(3x - 4)(x + 6)$

3. A bag holds 10 marbles: 4 red, 4 blue, and 2 yellow. Two marbles are drawn at random from the bag. To the nearest percent, what is the chance that one marble will be red and the other blue?

A. 18%

B. 20%

C. 25%

D. 36%

Write your answer in the box.

4. Eli has a collection of football and baseball cards. In all, he has 360 cards in his collection. Twice the number of baseball cards is 30 less than three times the number of football cards. How many more baseball cards than football cards does Eli have?

[ ] baseball cards

Select your answer from the choices.

5. The graph shows the number of children of each age enrolled in a class at a summer camp. There are 25 children in all. Find the mode, median, and mean of the ages of the children in the class.

Dino Camp for Young Scientists

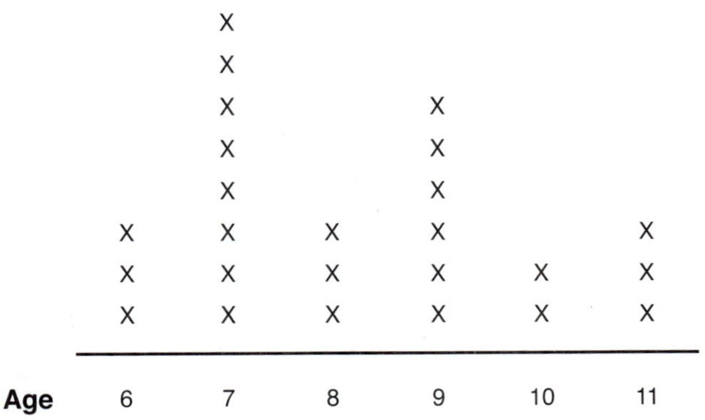

mode:

- 6
- 7
- 8
- 9

median:

- 6
- 7
- 8
- 9

mean:

- 6.5
- 7.6
- 8.2
- 9.4

6. A pet store has a large tank of tropical fish. The table shows the type of fish and the number of each in the tank.

Type of Fish	Number
Penguin Tetra	20
Gold Neon Tetra	10
Neon Jumbo Tetra	5
Cardinal Tetra	15

A customer buys 20 fish chosen randomly. Which of the following events is LEAST likely to happen?

A. The customer will get more Penguin Tetras than any other kind of fish.

B. The customer will get more Cardinal Tetras than Penguin Tetras.

C. The customer will get exactly five of each kind of fish.

D. The customer will get only two different kinds of fish.

Use the table and the graph to answer question 7. Write Year 1 or Year 2 in the box.

Year 1: Sales Data by Quarter

Quarter	Sales Revenue
1	$50,000
2	$45,000
3	$45,000
4	$35,000

Year 2: Sales Data by Quarter

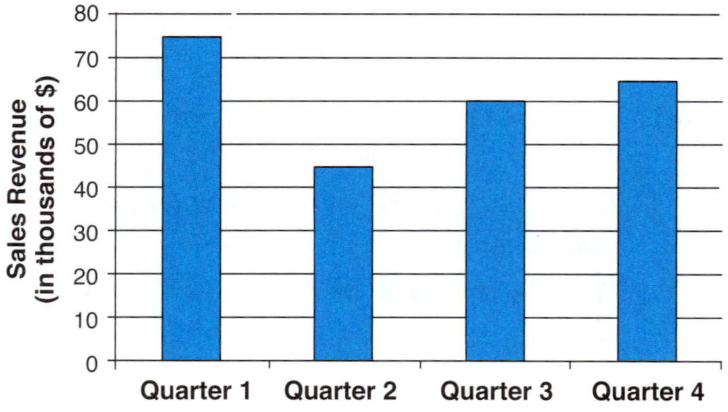

7. In which year was the difference in sales revenue between Quarter 1 and Quarter 4 the greatest?

Use the graph to answer question 8.

8. A quadratic equation of the form $ax^2 + bx + c = 0$ is graphed below.

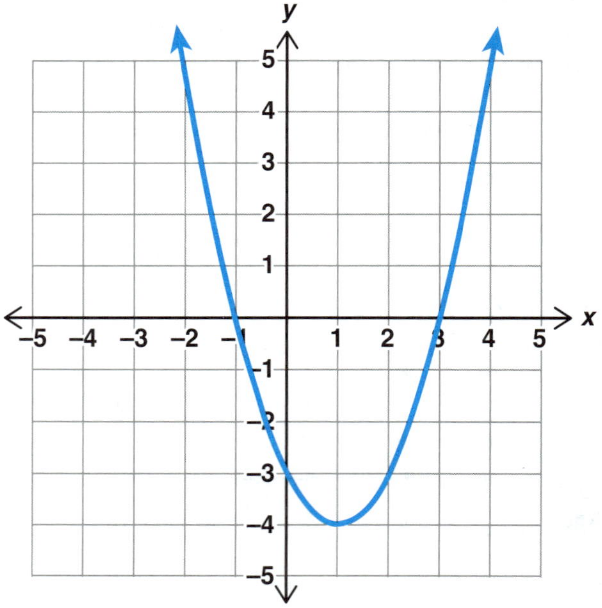

Based on the graph, what is one possible solution for x?

A. 4

B. 3

C. 2

D. 1

9. Which of the following completely simplifies the expression below?

$$\frac{n^2 - 6n - 7}{n^2 - 4n - 5}$$

A. $\frac{n-7}{n-5}$

B. $\frac{7}{5}$

C. $-5n - 6$

D. $2n + 2$

Write your answer in the box.

10. A large company gives its employees parking passes to display in their car windows. Each pass has 5 boxes with an arrangement of letters and numbers.

The first two boxes contain letters chosen from A, B, C, and D. The next three boxes contain numbers chosen from the digits 3, 4, 5, 6, 7, 8, and 9. The digits 0, 1, and 2 are not used.

If there are no repeating letters or numbers, how many different parking passes can the company make?

☐ parking passes

Use the scatter plot to answer question 11.

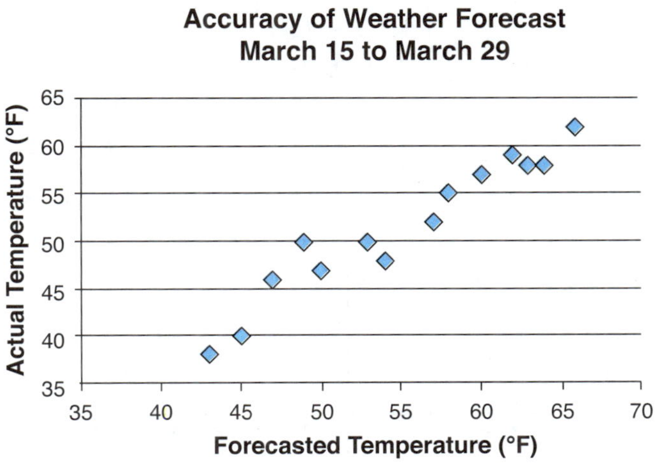

11. Based on the data in the scatter plot, if the forecasted temperature is 55°, which of the following choices is the best estimate for the actual temperature?

A. 45°

B. 51°

C. 55°

D. 57°

12. One of the equations in the following system is graphed on the coordinate plane below.

$$\begin{cases} x - y = 4 \\ 5x + 3y = 12 \end{cases}$$

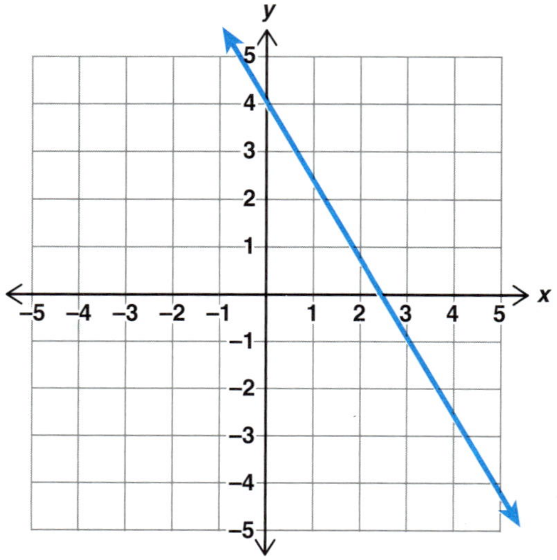

If the other equation is added to the graph, at which point will the equations intersect?

A. (0, 4)

B. (1, –3)

C. (3, –1)

D. (4, 0)

13. An order of 12 turkey club and 5 grilled cheese sandwiches costs \$121. An additional order of 6 turkey club and 8 grilled cheese sandwiches costs \$88. How much does one turkey club sandwich cost?

A. \$5.00

B. \$6.75

C. \$7.50

D. \$8.00

Write your answer in the box.

14. In a series, each new term is 4 less than the term that came before. The 5th term in the series is 18. What is the 15th term?

[ ]

15. What is the value of $f(-4)$ in the function $f(x) = 2x^2 - 7x + 23$?

[ ]

16. Multiply and simplify the expression:

$(3p - 5)(6p - 5)$

A. $18p^2 + 25$

B. $18p^2 - 15p - 25$

C. $18p^2 - 45p + 25$

D. $9p^2 - 20p - 10$

For questions 17–19, write your answer in the box.

17. Angela is graphing the following system on the coordinate plane below. To complete the graph, which region should she shade? Write the letter of the region in the box.

$$\begin{cases} y \geq x - 3 \\ y \leq -x + 1 \end{cases}$$

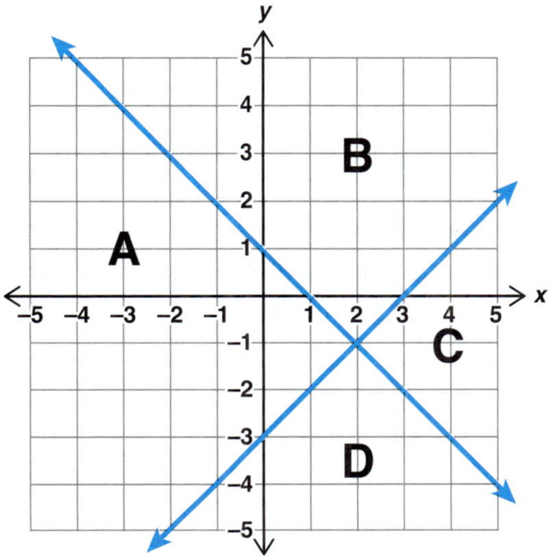

[ ]

18. Write the positive solution to this equation.

$7x^2 - 12 = 40x$

[ ]

19. Samantha has two job offers. Both jobs start at $17,000 per year. If she takes Job A, she will get a $300 raise at the end of each year. If she takes Job B, her salary (S) will be determined by a function based on how many years (t) she has worked for the company.

Function for Job B:

$$S(t) = 3.7t^2 + 335t + 17{,}000$$

After 4 years of work, which job offers Samantha the higher annual salary?

Job [ ]

20. The function g is shown on the graph below. In which interval does $g(x)$ have the least value?

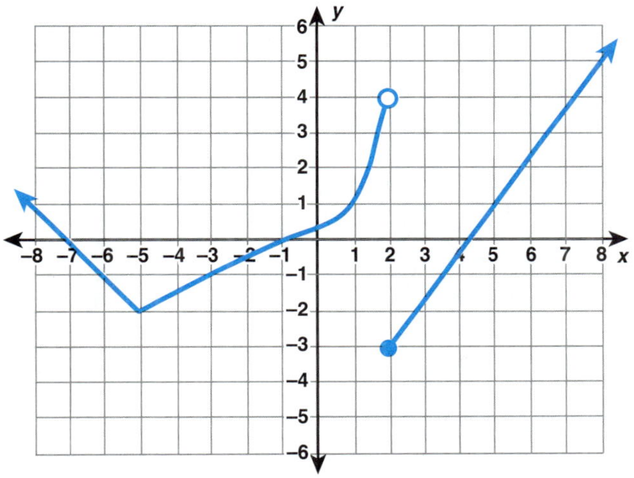

A. $x \leq -5$

B. $-5 < x \leq -1$

C. $-1 < x \leq 2$

D. $x > 2$

Select your answer from the choices.

21. Mr. Lopez gave his class of 30 students a pop quiz with 20 questions. Then he made the box plot below to show the results of the quiz. A score of 12 is a passing grade.

Pop Quiz Results

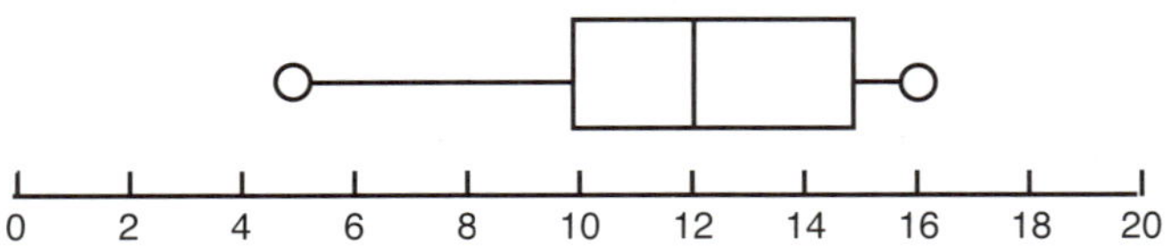

At least half the class passed the test.

- True
- False

A score of 11 is above the lower quartile.

- True
- False

Ten students scored above the upper quartile.

- True
- False

Use the graph to answer question 22.

Discretionary Spending
16- to 22-year-olds

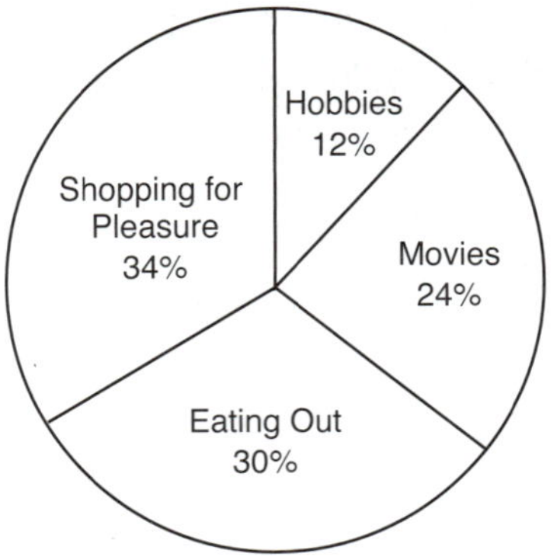

22. After she pays her bills each month, Lana has $300 left. Based on the graph, about how much will she spend on movies and eating out?

$ ______

23. Pro Sports is manufacturing a new style of baseball glove.

The equation $S = -\frac{3}{4}m^2 + 6m + 51$ can be used to model the number of sales in thousands (S) per month (m) for the first 12 months of production.

In which month will the sales drop to 36 thousand gloves?

A. Month 7

B. Month 8

C. Month 9

D. Month 10

Use the following data for questions 24 and 25.

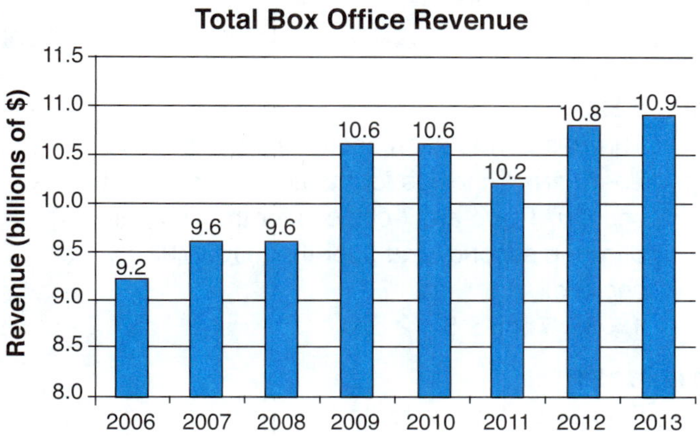

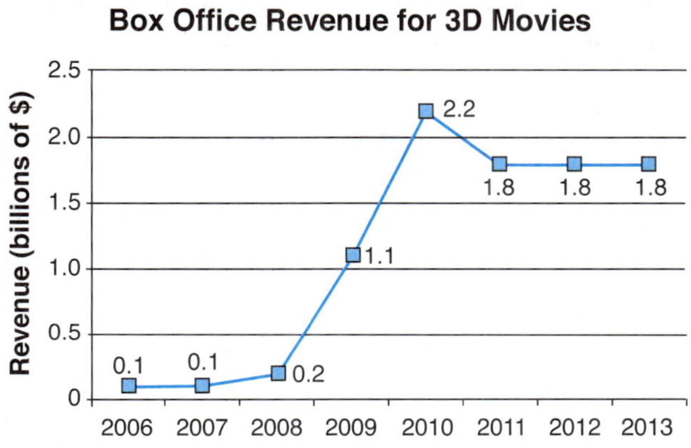

Source: Motion Picture Association of America, Inc.

24. In 2010, revenue from 3D movies was about what percent of the total box office revenue?

A. 8%

B. 21%

C. 48%

D. 79%

25. Consider the years 2011 through 2013. In each of these years, about fifty 3D movies were made. Based on the data for these years only, which of the following statements seems most accurate?

A. The percent of revenue generated by 3D movies is steadily increasing.

B. The percent of revenue generated by 3D movies is not changing.

C. The percent of revenue generated by 3D movies is declining.

Explain the reasoning for your choice in the space below.

Answers start on page 160.

Simulated Test Answer Key

1. **C. 28**
   The total of the bars to the left of 40 on the horizontal scale is 28: 2 + 4 + 9 + 7 + 6 = 28.

2. **B. $(3x - 8)(x + 3)$**
   Factor the trinomial, or use FOIL to check the answer choices.

3. **D. 36%**
   There is an 8 out of 10 chance of getting either red or blue on the first pick. The second pick must not be the same color, and it must not be yellow. There are 4 out of 9 choices that are desirable. Multiply the probabilities.
   $\frac{8}{10} \times \frac{4}{9} = \frac{32}{90} \approx 36\%$

4. **60 baseball cards**
   Write and solve a system.
   $\begin{cases} B + F = 360 \\ 2B = 3F - 30 \end{cases}$
   Rewrite the first equation as $B = 360 - F$ and substitute for B in the second equation.
   There are 210 baseball cards and 150 football cards.
   Subtract to find the difference: 210 − 150 = 60.

5. Mode: 7
   Median: 8
   Mean: 8.2
   $\frac{205}{25} = 8.2$

6. **C. The customer will get exactly five of each kind of fish.**
   You don't have to figure out the probabilities. Use your common sense about probability. The most restrictive choice has the least chance of happening. There are only 5 neon jumbo tetras. The chance of getting them all is very small.

7. **Year 1**
   In Year 1, the difference was $15,000.
   In Year 2, the difference was about $10,000.

8. **B. 3**
   The solutions are the values for x when the graph crosses the x-axis. The two solutions are −1 and 3.

9. **A. $\frac{n-7}{n-5}$**
   Factor both trinomials and simplify.
   $\frac{(n-7)\cancel{(n+1)}}{(n-5)\cancel{(n+1)}}$

10. **2,520**
    This is a permutation because order matters. There are 4 choices for the first letter and 3 for the second. There are 7 choices for the first number, 6 for the second, and 5 for the third. Multiply the choices.
    $4 \times 3 \times 7 \times 6 \times 5 = 2{,}520$

11. **B. 51°**
    Find 55 on the horizontal axis and then move upward. Estimate where the value would fall on an imaginary line drawn through the points on the graph. The point would be between 48 and 52. The best choice is 51°.

12. **C. (3, −1)**
    You can graph the line $x - y = 4$ to see where it will intersect the other line, or you can try each of the answer choices in the system to see which point makes *both* equations in the system true.

13. **D. $8.00**
    Write and solve a system.
    $\begin{cases} 12T + 5C = 121 \\ 6T + 8C = 88 \end{cases}$
    Multiply the second equation by −2 and add the equations.
    $$\begin{aligned} 12T + 5C &= 121 \\ -12T - 16C &= -176 \\ \hline -11C &= -55 \\ C &= 5 \end{aligned} \qquad \begin{aligned} 6T + 8(5) &= 88 \\ 6T &= 48 \\ T &= 8 \end{aligned}$$
    A cheese sandwich (C) costs $5, and a turkey sandwich (T) costs $8.

14. **−22**
    Work backward to figure out the first 5 terms. They are 34, 30, 26, 22, and 18. You know that 34 is the first term. Find the rule. The difference between the terms is −4. Add 38 to adjust the rule. The rule is $-4n + 38$.

n	1	2	3	4	5
$-4n$	−4	−8	−12	−16	−20
+38	34	30	26	22	18

    The 15th term is −4(15) + 38 = −22.

15. 83

Substitute –4 for x in the function and simplify.

$f(-4) = 2(-4)^2 - 7(-4) + 23$
$= 2(16) + 28 + 23$
$= 32 + 28 + 23$
$= 83$

16. C. $18p^2 - 45p + 25$

Use FOIL to multiply the binomials.

$(3p - 5)(6p - 5)$
$= 18p^2 - 15p - 30p + 25$
$= 18p^2 - 45p + 25$

17. Region A

Try a point from each region to see whether it will make both inequalities true.

Try (0, 0) from region A.

$0 \geq 0 - 3$	$0 \leq -0 + 1$
$0 \geq -3$ True	$0 \leq 1$ True

The origin (0, 0) makes both inequalities true, so you know that the origin is in the solution set. Angela should shade region A.

18. 6

This is a quadratic formula. You can solve it by factoring:

$7x^2 - 12 = 40x$
$7x^2 - 40x - 12 = 0$
$(7x + 2)(x - 6) = 0$
$x = -\frac{2}{7}$ and 6

The positive solution is 6.

19. Job B

At Job A, Samantha would receive 4 raises of \$300 each and then earn \$18,200 a year after 4 years.

\$17,000 + \$1,200 = \$18,200

Substitute 4 in the function to find her salary at Job B after 4 years.

$S(4) = 3.7(4)^2 + 335(4) + 17{,}000$
$= 59.2 + 1{,}340 + 17{,}000$
$= 18{,}399.20$

Job B is the better paying job.

20. C. $-1 < x \leq 2$

Look at the y-axis. The least value of $g(x)$ is at (2, –3), when $y = -3$. The function equals –3 when x equals 2, which is part of the interval $-1 < x \leq 2$.

21.

At least half the class passed the test.	**True**
A score of 11 is above the lower quartile.	**True**
Ten students scored above the upper quartile.	**False**

22. \$162

Movies + Eating Out is 24% + 30% = 54%.

54% of \$300 is \$162.

23. D. Month 10

This is a quadratic equation. Set S equal to 36 and solve.

$36 = -\frac{3}{4}m^2 + 6m + 51$
$0 = -\frac{3}{4}m^2 + 6m + 15$

You could use the quadratic formula.

$$\frac{-6 \pm \sqrt{6^2 - 4(-\frac{3}{4})(15)}}{2(-\frac{3}{4})}$$

$$= \frac{-6 \pm \sqrt{81}}{-\frac{3}{2}}$$

$$= \frac{-6 \pm 9}{-1.5}$$

$= -2$ or 10

Since you are given answer choices, you could also substitute the choices into the function to see which choice results in 36.

24. B. 21%

Find what percent 2.2 is of 10.6. Round to the nearest percent.

$\frac{2.2}{10.6} \approx 0.2075 \approx 21\%$

25. C. The percent of revenue generated by 3D movies is declining.

Box office revenue for 3D movies remained the same over the three years. However, total box office revenue increased during the same years. Over the 3-year period, the percent of revenue from 3D movies decreased.

Simulated Test **Evaluation Chart**

This simulated test measures how well you are able to apply the ideas taught in this book. The actual GED® Math Test will include additional topics that you can study in Books One and Two of this series.

On the chart below, circle the numbers of the problems that you answered incorrectly. Then review the skill area for each of those problems.

Problem Number	Skill Area	Unit
1, 5, 7, 11, 21, 22, 24, 25	Data Analysis	1
3, 6, 10	Counting and Probability	2
4, 12, 13, 17	Systems of Equations and Inequalities	3
14, 15, 19, 20	Functions	4
2, 9, 16	Polynomials and Rational Expressions	5
8, 18, 23	Quadratic Equations	6

Answer Key

UNIT 1

Working with Tables, pp. 18–19

1. **$22**
   Speaker sale price: $68
   DVD player sale price: $90
   $90 – $68 = $22
2. **Greater**
   Discount on HD television: $46
   Discount on DVD player: $30
3. **1,020 tons**
   2,100 – 1,080 = 1,020
4. March
5. March
6. **Logan County, by 2,060 tons**
   Carroll County: 2,250 – 520 = 1,730
   Logan County: 3,405 – 1,345 = 2,060
7. **4,449 miles**
   Mexico City to San Francisco: 1,883 miles
   San Francisco to New York: 2,566 miles
   1,883 + 2,566 = 4,449
8. The empty boxes represent the distance from each city to itself, which would be zero miles.

Bar Graphs and Histograms, pp. 20–21

1. Mac & Cheese, Pizza, and Grilled Cheese
2. **40**
   There were approximately 280 orders of pizza and 240 orders of lasagna.
   280 – 240 = 40
3. **$1,500**
   Year 1: $2,500
   Year 2: $4,000
   $4,000 – $2,500 = $1,500
4. Auction
5. **$3,000**
   Auction: about $5,000
   Food Booths: about $2,000
   $5,000 – $2,000 = $3,000
6. Petting Zoo
7. **Greater than 25 mpg**
   Greater than 25 mpg: 12 + 9 + 6 + 5 = 32
   Less than 25 mpg: 2 + 4 + 7 + 10 = 23
8. **55 cars**
   Add the number of cars for all intervals.
   2 + 4 + 7 + 10 + 12 + 9 + 6 + 5 = 55
9. **8 students**
   1 + 1 + 2 + 4 = 8
10. **73%**
    Number of students who passed: 10 + 8 + 4 = 22
    Total number of students: 1 + 1 + 2 + 4 + 22 = 30
    Percent: $\frac{22}{30} \approx 73\%$

Circle Graphs, pp. 22–23

1. **2,280 students**
   20% of 11,400 = 2,280
2. **Less than**
   Segments for Juniors, Seniors, and Super-Seniors add up to 35% of the student body. Freshmen represent 40%.
3. $0.05
4. **$1,400**
   $0.07 × $20,000 = $1,400
5. Volleyball, Soccer, and Tennis
6. 8%
7. **Car Repairs and Insurance *or* New Tires and Gas and Oil**
   A circle graph makes it easier to compare the parts to the whole.
8. Your answers should contain some of the information given here:
   A circle graph lets you easily compare segments or categories to each other and to the whole. A bar graph lets you easily compare categories to each other and find specific amounts.

Core Connections: Creating a Circle Graph, p. 23

	Votes	Percent	Degrees
Kohli	3,300	30%	108°
Sanderson	550	5%	18°
Morales	4,950	45%	162°
Souba	2,200	20%	72°

Scatter Plots, pp. 24–25

1. Yes, there is a negative correlation. As the driver's age increases, the maximum visibility decreases.
2. About 500 ft
3. 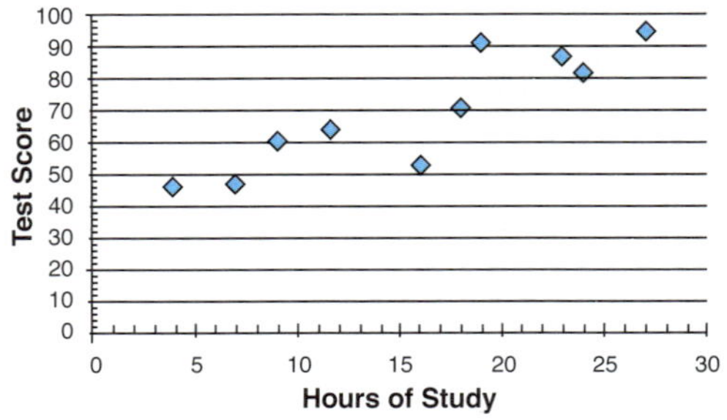

4. Yes, there is a positive correlation.
5. No correlation
6. Positive correlation
7. Negative correlation
8. Positive correlation

Line Graphs, pp. 26–27

1. 2006
2. **\$2,000 billion**
   2009: about \$14,000 billion
   2012: about \$16,000 billion
   \$16,000 billion – \$14,000 billion = \$2,000 billion
3. **2006**
   From 2005 to 2006, the gross national income increased about \$800.
4. 2009
5. \$0.40
6. **May to July**
   Gasoline prices stayed at \$3.60 per gallon for those three months.
7. March to April, September to November
8. **\$0.45**
   \$3.70 – \$3.25 = \$0.45
9. 1981
10. 2001
11. Germany
12. Japan's share of the market increased steadily from 1961 to 1981. It grew more slowly from 1981 to 1991 and began to decrease from 1991 to 2011.

Finding the Mean, pp. 28–29

1.	35	\$1.78	7
2.	168.6	7.4	\$48.50
3.	\$187	89	22.36
4.	5.7	\$234.20	7.6

5. \$78.50
6. 81
7. **14 miles**
   $\frac{3+8+8+0+4+5+x}{7} = 6$
   $28 + x = 42$
   $x = 14$
8. **3,069 books**
   $\frac{80+85+92+78+80+81+90+96}{8} = 85.25$
   $85.25 \times 36 = 3{,}069$
9. **No**
   60 days × 4 alarms = 240 alarms
10. No, it isn't possible for Carmel to receive an A. She would need to get a 106 on the last test, but the highest score she can receive is 100.
    $\frac{77+81+95+91+x}{5} = 90$
    $344 + x = 450$
    $x = 106$

Median and Mode, pp. 30–31

1.	\$23	22	11
2.	1.4	70	\$416
3.	18	22°	
4.	none	23 and 24	

5. Outlier: 3.2 kg
   Mean: 10.9 kg
   Median: 13.1 kg
6. Outlier: \$126,000
   Mean: \$91,000
   Median: \$85,000
7. Outlier: 44 in
   Mean: 61 in
   Median: 62.5 in
8. Outlier: 107
   Mean: 81
   Median: 76
9. **3**
   Average for 8 employees: 39
   Average for 9 employees: 42
   42 – 39 = 3
10. Answers may vary.

Using Typical Values, pp. 32–33

1. **B. median**
   The median is the middle score. Since an equal number of students scored above and below 76, 76 must be the middle score.
2. **C. mode**
   The mode is the value that occurs most often. The items that sell the best occur most often.

3. Answers may vary but should be similar to the sample answer:
The median is probably the most useful because it is the middle value, which means that half of the houses in the neighborhood would be cheaper than the median. If the median is near $150,000, you can probably afford a house in the neighborhood.

4. **A. A few people drink much fewer than 21 cups of coffee a week.**
If the mean is much lower than the median, then there are a few outliers that are very low.

5. **B. The mean is higher than the median.**
The 3 students that are 59 inches tall are outliers. These values are much higher than the rest of the data, so they will make the mean higher as well. The outliers won't have a large effect on the median value, however.

6. Answers may vary but should be similar to the sample answer:
Superstars in the NBA earn much higher salaries than $5 million a year, but most players make a much lower salary. The superstars are outliers, and they cause the mean to be much higher than the median.

7. **mean**
The word *average* usually refers to the mean.

8. You could conclude that most days in April were close to 40° and several days were much cooler than 51°.

9. Size 10

10. The median would also be a Size 10. Imagine that 100 dresses were sold. That would mean the boutique sold 15 Size 6's, 22 Size 8's, 48 Size 10's, and 15 Size 12's. If you listed the dresses sold in order of size, a Size 10 would be in the middle. Size 10 is the median value.

Weighted Averages, pp. 34–35

1. **74 mph**
$\frac{65(4) + 80(6)}{10} = 74$

2. **8.1**
$8(0.35) + 7.8(0.2) + 8.4(0.25) + 8.2(0.2) = 8.1$

3. **7.9 hours**
$\frac{3(3) + 6(18) + 8(54) + 11(15)}{90} \approx 7.9$

4. **66%**
$\frac{10(0.6) + 5(0.78)}{15} = 0.66 = 66\%$

5. **137.3**
$152(0.65) + 110(0.35) = 137.3$

6. **81**
$75(0.25) + 78(0.1) + 85(0.15) + 80(0.35) + 90(0.15) = 80.8 \approx 81$

7. **$160**
$\frac{2(\$175) + 3(\$150)}{5} = \$160$

8. **B. He batted about .200 the last month.**
$\frac{.300(3) + .200(1)}{4} = .275$

Mixed Review, pp. 36–37

1. About 15,000
2. Birds
3. About 10,000
4. Cats
5. 12 people
6. $\frac{7}{12}$
7. There is a positive correlation. The more resumes sent out, the more interviews received.
8. About 3
9. About 7,600
10. 2002
11. **About 7,363**
Estimate the values for 2007–2010:
2007: 7,650
2008: 7,600
2009: 7,300
2010: 6,900
Find the average: $\frac{7{,}650 + 7{,}600 + 7{,}300 + 6{,}900}{4} \approx 7{,}363$
12. **About 7,450**
Put the values from 2007 to 2010 in order:
6,900 7,300 7,600 7,650
Find the median:
$\frac{7{,}300 + 7{,}600}{2} = 7{,}450$
13. 20
14. 74
15. Longer than 3 minutes
16. **20%**
$\frac{15}{74} \approx 20\%$
17. Industry
18. $260,000
19. $\frac{1}{3}$
$33\% \approx \frac{1}{3}$
20. Mean: 9 oz
Median: 8 oz
Mode: 8 oz
21. **$2,000**
$\frac{2(\$800) + 5(\$1{,}500) + 4(\$2{,}750) + 1(\$3{,}900)}{12} = \$2{,}000$
22. Answers may vary but should include some of the following ideas:
The mean would not be very reliable because the 4 scores of 30 are outliers. The median would be the best typical value because the outliers would affect the median minimally, if at all.

Reading a Box Plot, pp. 38–39

1. lower quartile: 2.5
   upper quartile: 9.5
2. lower quartile: 4.95
   upper quartile: 5.9
3. lower quartile: 25
   upper quartile: 46
4. lower quartile: 31
   upper quartile: 89
5. 25
6. 40
7. 50%
8. **100**
   25% of the gym's members are between 25 and 40. 25% of 400 is 100.
9. Rumford Highway
10. 5 mph
11. On Shoreline Highway, 25% of cars drive between 60 and 70 mph. On Rumford Highway, 50% of cars drive between 60 and 70 mph.
    You cannot conclude that more cars drive between 60 and 70 mph on Rumford Highway. Shoreline Highway may have a higher total number of cars. If Shoreline has 1,000 cars, then 250 drive between 60 and 70 mph. If Rumford has 100 cars, then only 50 drive between 60 and 70 mph.
12.

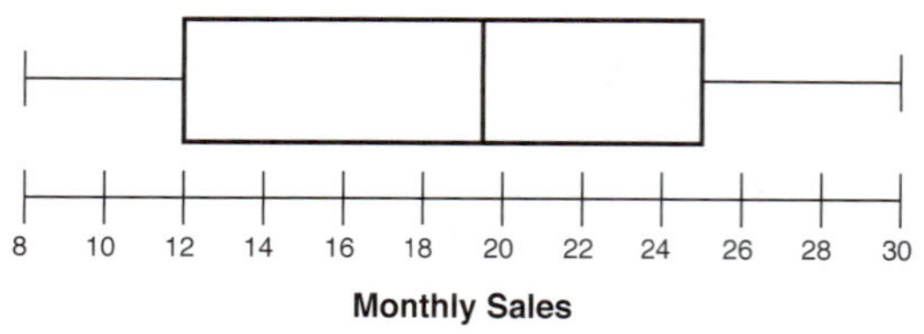

Distribution of Data, pp. 40–41

1. The data is skewed right.
2. **15–30 minutes**
   Most customer service phone calls take between 15 and 45 minutes.
3. The phone calls that take between 105 and 120 minutes are outliers.
4. Best Actress winners: 80 – 22 = 58
   Best Actor winners: 60 – 29 = 31
5. Best Actress
6. The Best Actress data set probably has outliers above the upper quartile (45–80). The maximum value is probably an outlier.
7. The data is skewed right.
8. Explanations may vary but should include the following ideas:
   The data for Best Actress is skewed right and has high variation. The data for Best Actor is symmetrical and has normal variation. The typical ages of Best Actress winners are lower than the typical ages of Best Actor winners, but the range for Best Actress winners is much greater than the range for Best Actor.
9. **Skewed left**
   When the mode is the highest and the mean is the lowest, data is skewed left.
10. Driving times through downtown during rush hour
11. **Skewed right**
    The salaries of the top players would create outliers on the right end of the scale. The data is skewed right.
12. **mean**
    When data is skewed right, the mean is the highest.

Making Predictions and Identifying Trends, pp. 42–43

1. The number of customers leaving increased from 2007 to 2011. The number of customers leaving decreased from 2011 to 2012 and remained steady in 2013.
2. The number of customers leaving has remained constant at 1.5 million during the last year.
3. Answers may vary but could include the following ideas:
   Since the number of customers leaving has decreased or remained steady the past 2 years, between 1 and 1.5 million customers probably left in 2014.
4. The number of R-rated tickets sold has mostly decreased since 2000. Over the last 3 years, sales have leveled off at around 200,000.
5. Answers may vary but could include the following ideas:
   Ticket sales have decreased slightly or remained steady over the last few years. That trend will probably continue.
6. Answers may vary but could include the following ideas:
   Since ticket sales for R-rated movies at this theater have declined and remain low, tickets for movies that are rated PG or G may sell better.
7. **a.** The price is increasing slowly.
   b. The price increased and then leveled off.
   c. The price has decreased since February.
8. Answers may vary. A line graph would probably be most useful.

9. RAO: around 2.27
   LEC: around 5.46
   KBW: around 10.24
10. Answers may vary but should include the following ideas:
    No, you couldn't use the table for the rest of the year. This table represents only 5 months. It doesn't give enough information to predict the remaining 7 months of the year.

Understanding Correlation, pp. 44–45

1. Car weight and average mileage have a negative linear correlation.
2. 20 mpg
3. Below
4. Positive: 1–10 hours
   Negative: greater than 10 hours
5. About 10 hours
6. Answers may vary. The following is a sample answer: There are other factors involved in getting a good grade. Usually, you should get a good night's sleep before taking a test. Students studying more than 10 hours might be studying too late at night and hurting their test score.

Core Connections: Correlation and Causation, p. 45

1. No causation
2. Causation
3. Causation
4. No causation

Sampling a Population, pp. 46–47

1. Sample: work done between 2:30 and 3:30
   Population: work done throughout the day
   Representative? No. The supervisor should check at different times during the day.
2. Sample: 2,000 surveyed owners
   Population: all new pickup owners
   Representative? Yes. The manufacturer surveys a variety of owners randomly.
3. Sample: residents of every 5th apartment
   Population: all residents in the complex
   Representative? Yes. Residents were randomly chosen.
4. Sample: one package of cookies
   Population: all packaged cookies
   Representative? No. The size of the sample is too small.
5. **D. The sample is not representative.**
   By surveying only band members, the survey is not a random, unbiased representation of all students at the high school. While the sample size is probably adequate, it is not representative.
6. Answers may vary. The following is a sample answer: Compare prices of about 20 varied food items found in Value Supermarket and at least 2 other stores.
7. **C. 100 freshmen in a single dormitory**
   The dormitory is likely to have a mixture of students in different studies from different hometowns.
8. Answers may vary. The following is a sample answer: Choose about 40 people of different ages, male and female, who swim at different times during the day.

Using Two Data Sources, pp. 48–49

1. **$142.56**
   Total distance: 985 mi
   Mileage: 28 mpg
   $985 \div 28 \approx 35.2$ gallons
   $35.2 \times \$4.05 = \$142.56$
2. **$18.55**
   Total distance: 520 mi
   Compact mileage: 22 mpg
   $520 \div 22 \approx 23.6$ gallons
   $23.6 \times \$3.50 = \$82.60$

   SUV mileage: 18 mpg
   $520 \div 18 \approx 28.9$ gallons
   $28.9 \times \$3.50 = \$101.15$

   $\$101.15 - \$82.60 = \$18.55$
3. **969,784 people**
   54.2% of $1{,}789{,}270 \approx 969{,}784$
4. **216,501 or 216,502 people**
   $54.2\% - 42.1\% = 12.1\%$
   12.1% of $1{,}789{,}270 \approx 216{,}502$
5. **larger**
   The bar graph shows that Obama received slightly more than 50% of the vote in the country, but he received 54.2% of the vote in Oregon.
6. **5%**
   $47\% - 42\% = 5\%$
7. July and August
8. **No**
   The month with the lowest temperature is January. The months with the fewest accidents are March and April.
9. July and August

10. Answers may vary but could include some of the following ideas:
Temperatures are highest in the summer, and the summer months generally have more accidents. However, the months with the lowest temperatures don't always have fewer accidents. In the summer, it could be that there are more accidents because students are out of school or people may be out more in nicer weather.

Drawing Conclusions, pp. 50–51

1. Not justified
2. Not justified
3. Justified
4. This graph is misleading because the basketball looks 10 times as large as the baseball, but basketball ticket prices are a little more than twice as expensive. This graph would be less misleading if it were a simple bar graph.
5. Not justified
6. Justified
7. Justified
8. This graph is misleading because the vertical scale starts at 40%, not 0%. It would be less misleading if the scale were adjusted and also changed to show increments of 5% instead of 2%.
9. Wilson
10. **12%**
$2.12 - 1.87 = 0.25$
$\frac{0.25}{2.12} \approx 12\%$
11. Answers may vary. Sample answers could include:
Elementary school budgets in Hodgson County decreased by about 10% in Year 2.
or
Elementary school budgets in Hodgson County decreased by about $240,000 in Year 2.
12. Answers may vary but could include the following ideas:
The vertical scale on this graph doesn't start at 0, which makes it slightly misleading. However, since the data values are so close together, the graph might be difficult to read if you adjust the scale.

Unit 1 Review, pp. 52–55

1. $128 - 68 =$ **60**
2. 91 km
3. 50%
4. 100 km
5. B. skewed right
6. **47%**
Number of dogs 2 or older:
$10 + 9 + 4 + 4 + 3 + 2 = 32$
Total number of dogs: $21 + 15 + 32 = 68$
$\frac{32}{68} \approx 47\%$
7. positive correlation
8. no correlation
9. negative correlation
10. positive correlation
11. A. Above the line of best fit
12. $8,000
13. **B. $672,000**
12% of 200 = 24
24 × $28,000 = $672,000
14. **Sales**
Supervisory: 10% of 200 = 20
20 × $46,000 = $920,000
Sales: 22% of 200 = 44
44 × $34,000 = $1,496,000
15. **C. 5**
$\frac{1+2+5+8+4+2+4+4+10+1+3+0+6+5+x}{15} = 4$
$\frac{55+x}{15} = 4$
$55 + x = 60$
$x = 5$
16. **A. 500**
Let x be the mean number of miles for the 20 employees.
$\frac{20x + 3{,}398}{21} = 638$
$20x + 3{,}398 = 13{,}398$
$20x = 10{,}000$
$x = 500$
17. **C. 4,800**
$\frac{960}{3{,}000} = 32\%$
32% of 15,000 = 4,800
18. **12 students**
The median is the middle value. If there are 25 students in the class, 12 must score higher than 73, and 12 must score lower than 73.
19. **C. 7.6**
$7(0.15) + 9(0.35) + 6(0.4) + 10(0.1) = 7.6$
20. $2
21. **B. $2**
There are 27 items in the bin. The middle value must be the 14th value. Since there are 6 books for $1 and 9 books for $2, the 14th value must be $2.
22. No, this conclusion is not justified. Only the last bar represents 5 years. You can't compare the lengths of the bars because they represent different periods of time.
23. C. 3,500

24. D. 1992 and 1996

25. C. 2000

26. D. The number has remained fairly constant.

UNIT 2

Counting Strategies, pp. 58–59

1. **24**
   Try making an organized list or a tree diagram. Remember that the two sitcoms are different shows.
2. **6**
   Draw a picture. The mother is always in the same seat, so you are really counting how many different ways the 2 children could sit in the remaining 3 seats.
3. **15**
   There are 5 colors for each of the 3 styles of shirts: $5 \times 3 = 15$.
4. **15**
   Make a list or connect the dots and count carefully. You could also think: A connects to 5 other dots, B connects to 4 other dots, C connects to 3, D connects to 2, and E connects to 1.
   $5 + 4 + 3 + 2 + 1 = 15$
5. **20**
   For each of the 5 first actions, there are 4 possible second actions: $5 \times 4 = 20$.
6. **10**
   Draw the different routes. You can also make an organized list. Think: It will take 5 steps to get to the house: 2 will go north and 3 will go east. Write the list showing all the possible placements for N (north).

Permutations, pp. 60–61

1. **336**
   $8 \times 7 \times 6 = 336$
2. **5,850**
   $26 \times 25 \times 9 = 5{,}850$
3. **11,880**
   $12 \times 11 \times 10 \times 9 = 11{,}880$
4. **120**
   $5 \times 4 \times 3 \times 2 \times 1 = 120$

Core Connections: Circular Permutations, p. 61

1. **24**
   $(5 \times 4 \times 3 \times 2 \times 1) \div 5 = 24$
2. **120**
   $(6 \times 5 \times 4 \times 3 \times 2 \times 1) \div 6 = 120$

Combinations, pp. 62–63

1. **C** It doesn't matter in which order the balls are drawn.
2. **P** The order the students are chosen will decide who is president and who is secretary. Order matters.
3. **P** The order of the digits matters in a telephone number.
4. **C** The contents of the sundae are the same no matter in what order the scoops are placed in the bowl.
5. **P** The order of the players in the lineup matters.
6. **C** The players don't start the game in any order. They are simply the five starters.
7. **1,365 committees**
   $\frac{15 \times 14 \times 13 \times 12}{4 \times 3 \times 2 \times 1} = \frac{32{,}760}{24} = 1{,}365$
8. **15 groups of 2 T-shirts**
   $\frac{6 \times 5}{2 \times 1} = \frac{30}{2} = 15$
9. **816 study groups**
   $\frac{18 \times 17 \times 16}{3 \times 2 \times 1} = \frac{4{,}896}{6} = 816$
10. **20 combinations**
    $\frac{6 \times 5 \times 4}{3 \times 2 \times 1} = \frac{120}{6} = 20$
11. **1,326 pairings**
    $\frac{52 \times 51}{2 \times 1} = \frac{2{,}652}{2} = 1{,}326$
12. **142,506 groups**
    $\frac{30 \times 29 \times 28 \times 27 \times 26}{5 \times 4 \times 3 \times 2 \times 1} = \frac{17{,}100{,}720}{120} = 142{,}506$
13. a. **1,140 groups**
    $\frac{20 \times 19 \times 18}{3 \times 2 \times 1} = \frac{6{,}840}{6} = 1{,}140$
    b. **120 ways**
    $5 \times 4 \times 3 \times 2 \times 1 = 120$
14. **6 ways**
    Since Matt is going to swim last, you need to consider only the first 3 swimmers: $3 \times 2 \times 1 = 6$ ways.
15. **216 codes**
    Think about the choices for each digit. There are 9 choices for the first digit, 3 for the second digit, and 8 for the third digit. Multiply: $9 \times 3 \times 8 = 216$
16. **12 orders**
    You could make a list.
    C: circle, S: star, B: block, H: half
    Then count the choices that do not have a B in the 1st or 4th position. There are 12.

CSBH	SCBH	HSBC
CBSH	SBCH	HBSC
CHBS	SHBC	HCBS
CBHS	SBHC	HBCS

Another method is to use the idea that choices multiply:
Think about the choices for each space in the row. You can't use the completely shaded block in the 1st space, so there are 3 choices for the 1st space. Now consider the 4th space. There are 2 possible choices left for the 4th space. Imagine that you have placed 2 blocks. You have 2 choices left for the 2nd space and 1 choice left for the 3rd space. Multiply the choices: $3 \times 2 \times 1 \times 2 = 12$

Simple Probability, pp. 64–65

1. $\frac{3}{8}$
   Three out of 8 numbers are greater than 12.
2. **25%**
   Two out of 8 numbers are 20 or 25.
   $\frac{2}{8} = \frac{1}{4} = 25\%$
3. $\frac{5}{8}$
   Three out of 8 numbers are 10, so 5 out of 8 numbers are not 10.
4. $\frac{1}{4}$
   Thirteen out of 52 cards are diamonds, which equals 1 out of 4.
   $\frac{13}{52} = \frac{1}{4}$
5. $\frac{1}{13}$
   There are 4 aces out of 52 cards.
   $\frac{4}{52} = \frac{1}{13}$
6. $\frac{2}{13}$
   There are four 5s and four 6s.
   $\frac{8}{52} = \frac{2}{13}$
7. **1 to 3**
   One out of every 4 cards is a heart, so 3 out of every 4 cards are not. The odds in favor of drawing a card this is a heart are 1 to 3.
8. **50%**
   Three out of 6 numbers are even.
   $\frac{3}{6} = \frac{1}{2} = 50\%$
9. **1 to 2**
   Out of 6 numbers, 2 numbers are divisible by 3, and 4 numbers are not. The odds are 2 to 4, or 1 to 2.
10. $\frac{5}{6}$
    The probability of rolling a 1 is $\frac{1}{6}$, so the probability of *not* rolling a 1 is $\frac{5}{6}$.
11. $\frac{1}{45}$
    The number of 2-person combinations is $\frac{10 \times 9}{2 \times 1} = \frac{90}{2}$ = 45. Only 1 of those combinations includes both Candice and Lior.
12. a. **8 toppings**
    b. If there are 8 toppings, the number of permutations of toppings is 8 × 7 = 56. But the order of the toppings doesn't matter, so divide by 2 × 1 = 2. There are 56 ÷ 2 = 28 combinations. One of those combinations is pepperoni and pineapple.

Compound Probability, pp. 66–67

1. $\frac{1}{8}$
   $\frac{1}{2} \times \frac{1}{2} \times \frac{1}{2} = \frac{1}{8}$
2. $\frac{1}{36}$
   $\frac{1}{6} \times \frac{1}{6} = \frac{1}{36}$
3. $\frac{9}{38}$
   $\frac{10}{20} \times \frac{9}{19} = \frac{1}{2} \times \frac{9}{19} = \frac{9}{38}$
4. $\frac{1}{3}$
   $\frac{1}{6} + \frac{1}{6} = \frac{2}{6} = \frac{1}{3}$
5. $\frac{1}{10}$
   There is a 2 out of 5 chance of getting math for the first class and a 1 out of 4 chance of getting math for the second class.
   $\frac{2}{5} \times \frac{1}{4} = \frac{2}{20} = \frac{1}{10}$
6. $\frac{1}{11}$
   $\frac{6}{12} \times \frac{5}{11} \times \frac{4}{10} = \frac{1}{2} \times \frac{\cancel{5}^{1}}{11} \times \frac{4}{\cancel{10}_{2}} = \frac{4}{44} = \frac{1}{11}$
7. **10%**
   $\frac{10}{300} + \frac{20}{300} = \frac{30}{300} = \frac{1}{10} = 10\%$
8. $\frac{3}{8}$
   You could make a list. Using B for boy and G for girl, the possibilities are BBB, BBG, BGB, BGG, GGG, GGB, GBG, GBB. Three out of the 8 possibilities have 2 boys and 1 girl.
   Alternate method: There is a 50% chance of having either a boy or a girl—like flipping a coin. The chance of getting any combination is $\frac{1}{2} \times \frac{1}{2} \times \frac{1}{2} = \frac{1}{8}$. There are 3 possible orders that have 2 boys and 1 girl: BBG, BGB, and GBB.
   $\frac{1}{8} + \frac{1}{8} + \frac{1}{8} = \frac{3}{8}$.

Using Probability for Predictions, pp. 68–69

Your answers may vary somewhat since you may apply different reasoning. If you aren't sure that your answer makes sense, explain your thinking to a partner or a friend.

1. Between 64% and 70%; probably closer to 64%.
   Last year, the basketball player made 64% of his free throws. This year, he is off to a good start, making 70% of his free throws. Perhaps he has improved, but it is more likely that his rate will level off as the season goes on.
2. a. **58%**
   290 ÷ 500 = 0.58 = 58%
   b. **42 voters**
   If 58% support Hinckley, then 100% – 58% = 42% support Kelly. 42% of 100 is 42 voters.
   c. **D. Kelly**
   Almost 47% of Hinckley's supporters are women. About 54% of Kelly's supporters are women.
3. Your answers may vary.
   An answer close to 22 is the best choice. Even though 4 out of the last 10 cars were white, the sample is too small to be predictive.

4. Your answers may vary.
   7 or 8 is a good prediction. Your answer should be about $\frac{1}{3}$ of your answer to problem 3.

Core Connections: Area and Probability, p. 69

1. **32%**
   There are 8 shaded squares out of 25 squares in all.
   $\frac{8}{25} = \frac{32}{100} = 32\%$
2. **1 to 3**
   Together, red and blue make up 25% of the spinner. The odds compare 25% to the chance of not getting red or blue: 75%. $\frac{25}{75} = \frac{1}{3}$

Experimental Probability, pp. 70–71

1. **45%**
   9 of the 20 trials are equal to or greater than 10.
   $\frac{9}{20} = 45\%$
2. **5**
   Only 2 out of 20 trials, or 10%, are less than 5. 10% of 50 is 5.
3. Spinner B
4. Spinner C
5. Spinner A
6. Spinner A should yield the lowest scores even though you can't get a 1 on Spinner A. Consider the opposite question: Which spinners will produce the greatest scores? Analyze the space for the numbers 3 and 4. Spinner B gives $\frac{3}{4}$ of its space to 3 and 4. Spinner C gives more than $\frac{3}{4}$ of its space to 3 and 4. One-half of Spinner A is used for 3 and 4. Since B and C give the most space to the greater numbers, they are more likely to produce high numbers.
7. $\frac{3}{20}$ or 15%
8. $\frac{4}{20}$ or 20%
9. The chance of getting 3 red cards is $\frac{10}{30} \times \frac{9}{29} \times \frac{8}{28} = \frac{720}{24360} \approx 3\%$. Since the experiment has only 20 trials, it is reasonable that no trial yielded 3 red. If you continued the experiment, you would eventually get 3 red.

Unit 2 Review, pp. 72–73

1. **160 cupcakes**
   Multiply the options.
   $4 \times 8 \times 5 = 160$
2. **14 diagonals**
   Each vertex connects to 4 other vertices. For example, *A* connects to *C*, *D*, *E*, and *F*. There are 7 vertices, so you can calculate $7 \times 4 = 28$. But 28 isn't the answer. Connecting *A* to *C* is the same as connecting *C* to *A*. When you multiplied, you counted each diagonal twice.
   $28 \div 2 = 14$ diagonals
3. **A. 5%**
   The only number common to both stacks is 4. There is a $\frac{1}{5}$ probability of drawing a 4 in the first stack and $\frac{1}{4}$ of drawing a 4 in the second stack. To find the probability of drawing a 4 in both stacks, multiply.
   $\frac{1}{5} \times \frac{1}{4} = \frac{1}{20} = 5\%$
4. **3,024 finishes**
   This is a permutation because the order in which the runners finish the race matters.
   $9 \times 8 \times 7 \times 6 = 3{,}024$ finishes
5. **20%**
   6 out of 30 total jelly beans are pink. $\frac{6}{30} = \frac{1}{5} = 20\%$
6. **A. 2 to 3**
   12 of the 30 are red; 18 are not red. The odds are 12 to 18, or 2 to 3.
7. **C. $\frac{2}{3}$**
   This is an either/or probability problem, so you need to add probabilities: $\frac{6}{30} + \frac{4}{30} = \frac{10}{30} = \frac{1}{3}$. There is a $\frac{1}{3}$ probability of getting pink or black, so there is a $1 - \frac{1}{3} = \frac{2}{3}$ probability of not getting pink or black.
8. **924 groups**
   This is a combination problem because the order in which the players kneel doesn't matter: ABCDEF is the same group as FEDCBA.
   Find the permutations of 6 out of 12:
   $12 \times 11 \times 10 \times 9 \times 8 \times 7 = 665{,}280$
   Find the permutations of 6 players:
   $6 \times 5 \times 4 \times 3 \times 2 \times 1 = 720$.
   Divide: $665{,}280 \div 720 = 924$
9. **C. Choosing D or I**
   Evaluate the choices.
   V or E: There are 2 cards.
   Not D or I: There are 2 cards.
   D or I: There are 4 cards.
   Not I or E: There are 3 cards.
   Choosing D or I has the greatest chance of happening.
10. **D. $\frac{1}{3}$**
    There are 36 possible dice rolls ($6 \times 6 = 36$).

Ten possibilities have exactly one 5:		Two possibilities add to make 3:
5, 1	6, 5	2, 1
5, 2	4, 5	1, 2
5, 3	3, 5	
5, 4	2, 5	
5, 6	1, 5	

$\frac{10}{36} + \frac{2}{36} = \frac{12}{36} = \frac{1}{3}$

11. **B. 40%**
    The team scored 4 runs in exactly 5 games. It won 2 of those games. $\frac{2}{5} = 40\%$

12. C. 65
Out of 25 games, the team scored 6 runs or more in 10 games: $\frac{10}{25} = 40\%$. 40% of 162 is about 65 games.

13. C. 36%
The probability of raining in Cedar Break is $\frac{3}{5}$.
The probability of not raining in Shadow Falls is $\frac{3}{5}$.
The probability of both events is $\frac{3}{5} \times \frac{3}{5} = \frac{9}{25} = 36\%$.

14. Event A
The probability that it will rain in both towns is $\frac{3}{5} \times \frac{2}{5} = \frac{6}{25}$. The probability that it will rain in neither is $\frac{3}{5} \times \frac{2}{5} = \frac{6}{25}$. So the probability of Event B, both or neither, is $\frac{6}{25} + \frac{6}{25} = \frac{12}{25}$. The only other possibility is that it rains in exactly one town (Event A): $1 - \frac{12}{25} = \frac{13}{25}$. Event A has a slightly greater chance of occurring.

UNIT 3

Systems of Equations, pp. 76–77

1. (–1, –3)

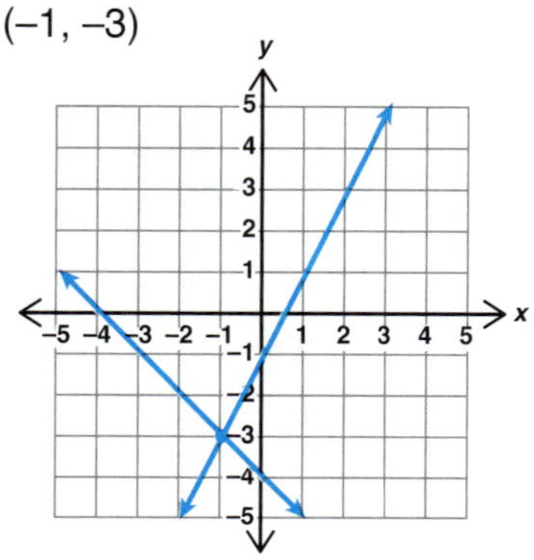

2. (–1, 2)

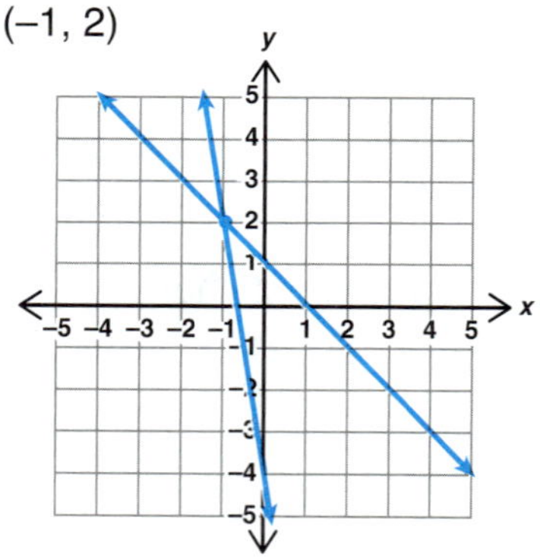

3. None

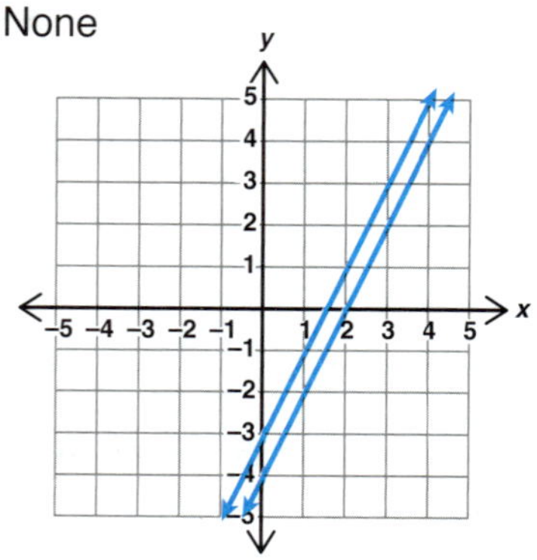

4. (4, 1)

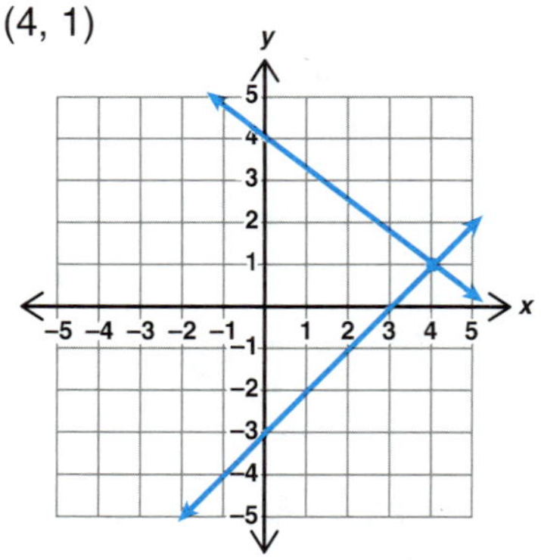

5. Infinite

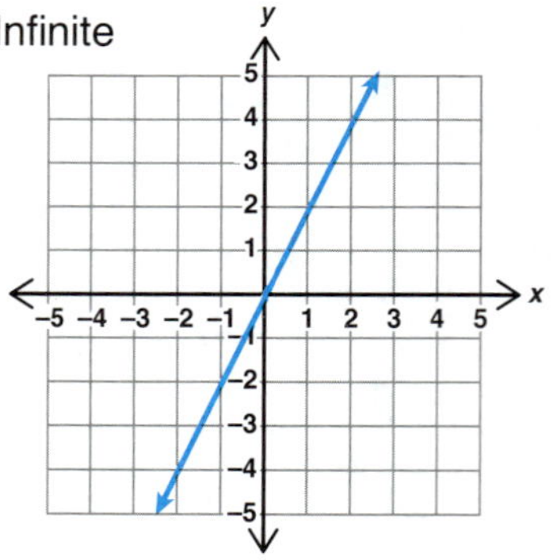

6. (3,–4)

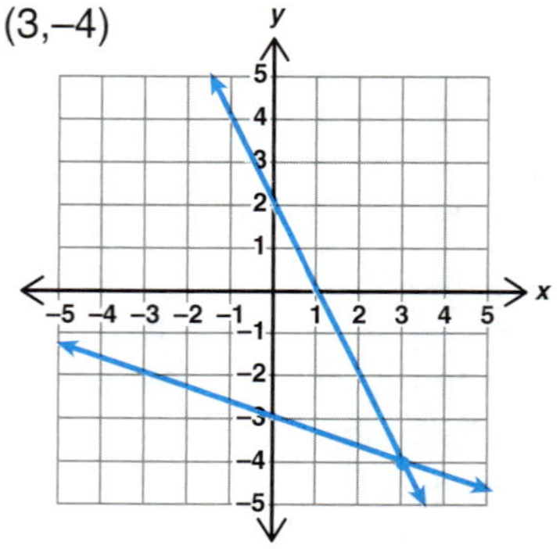

Solving Systems Using Substitution, pp. 78–79

For problems 1–5, one equation in the system has an isolated variable. Substitute the expression equal to that variable into the other equation.

1. (0, 4)
2. (–3, –1)
3. (–2, 4)
4. (–3, 0)
5. (0, 2)

For problems 6–10, first isolate a variable in one of the equations. Then substitute.

6. (–8, 6)
Isolate x. Rewrite $x + 4y = 16$ as $x = -4y + 16$.

7. (3, –8)
Isolate x. Rewrite $x - 2y = 19$ as $x = 2y + 19$.

8. (4, 2)
Isolate y. Rewrite $-3x + y = -10$ as $y = 3x - 10$.

9. (–1, 0)
Isolate y. Rewrite $-8x + y = 8$ as $y = 8x + 8$.

10. (–6, –1)
Isolate x. Rewrite $x - 6y = 0$ as $x = 6y$.

Solving Systems Using Elimination, pp. 80–81

1. (4, 6)
To eliminate y, multiply $7x - 3y = 10$ by –2 to get $-14x + 6y = -20$. Then add the equations.

2. (3, –2)
Isolate x. Rewrite $x + 3y = -3$ as $x = -3y - 3$.

3. (7, –4)
To eliminate x, multiply $-2x - 9y = 22$ by 3 to get $-6x - 27y = 66$. Then add the equations.

4. (–4, 0)
To eliminate y, multiply $3x - 11y = -12$ by 2 to get $6x - 22y = -24$. Then add the equations.

5. (–3, –3)
Isolate y. Rewrite $-2x + y = 3$ as $y = 2x + 3$.

6. (–4, –6)
To eliminate y, multiply the first equation by 3 and the second equation by –2. (You could also multiply by –3 and 2.) Then add the equations.

7. (–2, 4)
To eliminate y, multiply the first equation by 5 and the second equation by 2. Then add the equations.

8. (–7, –9)
To eliminate x, multiply the first equation by 10 and the second equation by –3. (You could also multiply by –10 and 3.) Then add the equations.

9. (9, 3)
To eliminate x, multiply the first equation by 5 and the second equation by 2. Then add the equations.

10. (–2, 2)
To eliminate y, multiply the first equation by 5 and the second equation by –2. (You could also multiply by –5 and 2.) Then add the equations.

Write Systems to Solve Word Problems, pp. 82–83

1. The missing sides measure 8 and 24.

2. 12 dimes and 19 quarters

3. The Tigers scored 117 and the Bulldogs scored 92 points.
Let T = the Tigers' points and B = the Bulldogs' points. The system can be written as:

$$\begin{cases} T + B = 209 \\ T = B + 25 \end{cases}$$

4. Ben had 18 misses and 22 hits.
Let m represent misses and h represent hits. The system can be written as:

$$\begin{cases} h + m = 40 \\ 10h + (-15m) = -50 \end{cases}$$

5. 36 grams
Let t represent trapezoids and p represent pentagons. The system can be written as:

$$\begin{cases} 2t + p = 42 \\ 5t = 8p \end{cases}$$

The variable $t = 16$, and the variable $p = 10$. The question asks you to find the value of $2p + t$, which equals $2(10) + 16 = 36$ grams.

6. Cathy is 27 years old.
Let b represent Bob's age now. Let c represent Cathy's age now. The system can be written as:

$$\begin{cases} b + c = 84 \\ b + 3 = 2(c + 3) \end{cases}$$

7. The number is 36.
Let t represent the tens place and o represent the ones place. The system can be written as:

$$\begin{cases} 10t + o = 4(t + o) \\ o = t + 3 \end{cases}$$

8. The fraction is $\frac{18}{30}$.
Let n represent the numerator and d represent the denominator. The system can be written as:

$$\begin{cases} n = d - 12 \\ \frac{n}{d} = \frac{3}{5} \end{cases}$$

Graph Linear Inequalities, pp. 84–85

	Point	Graph A	Graph B
1.	(0, 0)	✓	
2.	(1, 3)	✓	
3.	(–3, –4)	✓	✓
4.	(–1, 0)	✓	
5.	(4, –2)		✓

6.

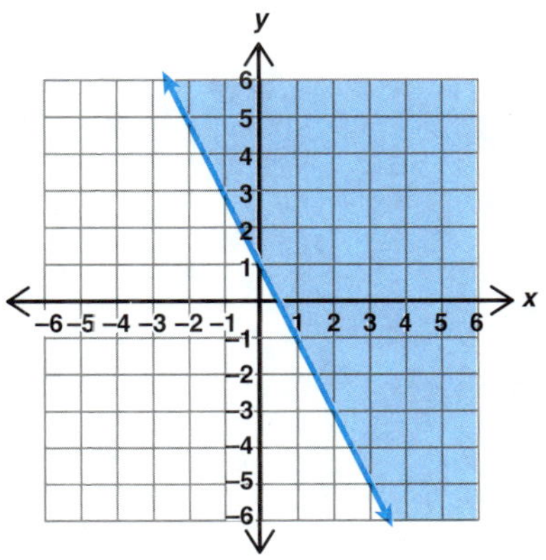

7.

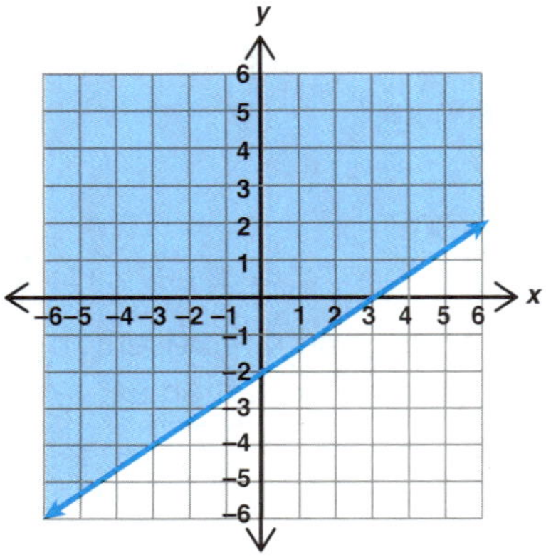

8.

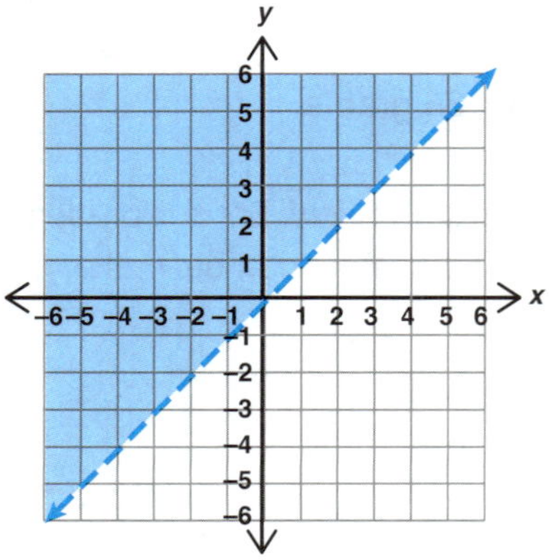

9. Your graph should look something like this:

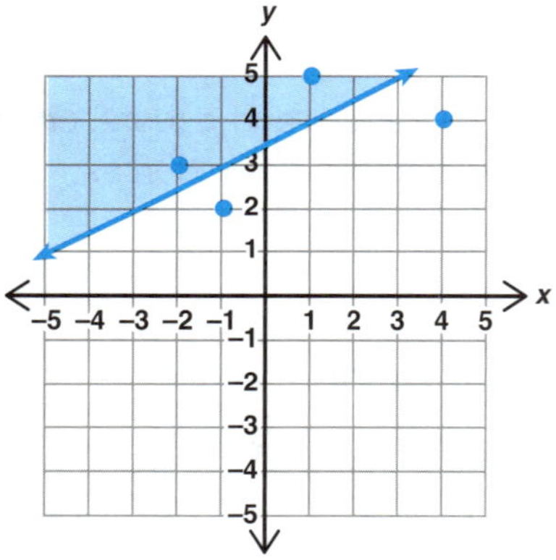

Solve Systems of Linear Inequalities, pp. 86–87

1.

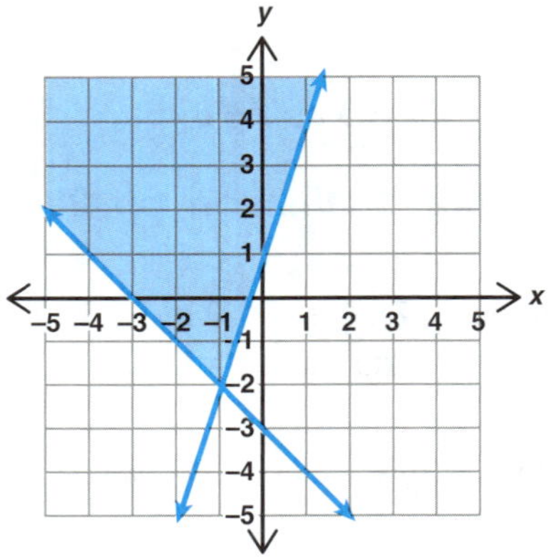

2.

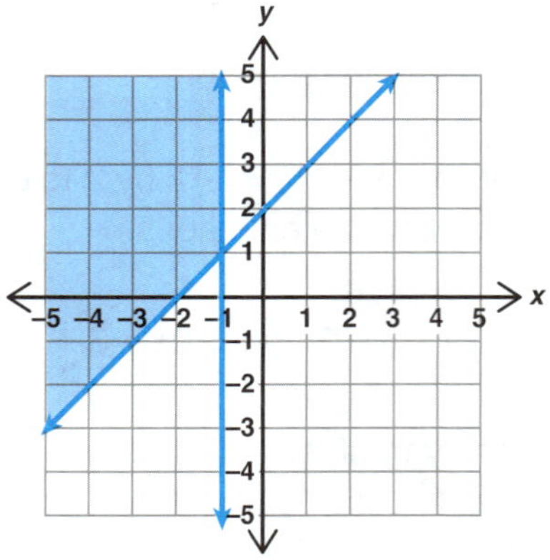

3.

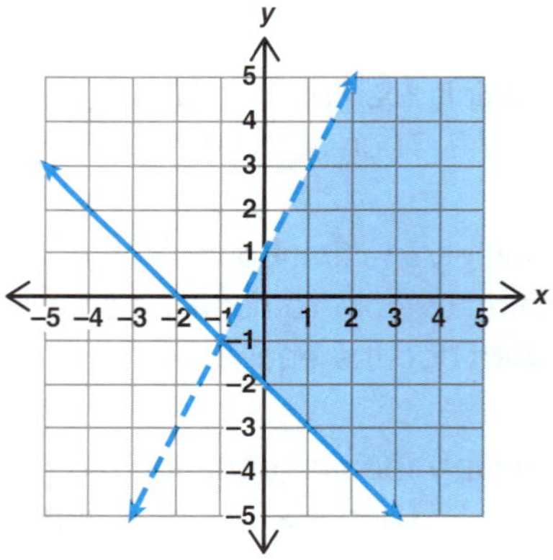

4.

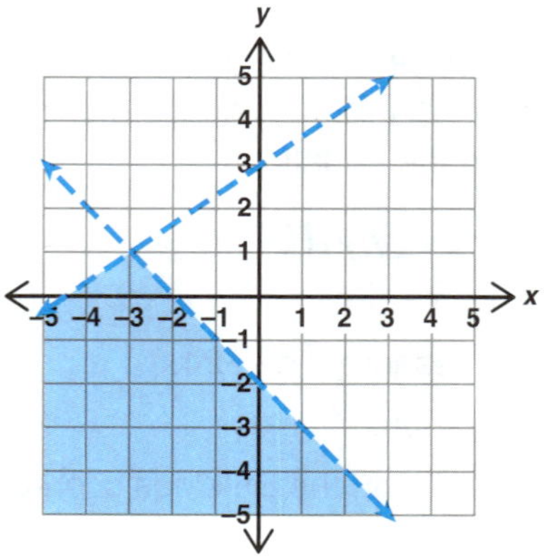

5.

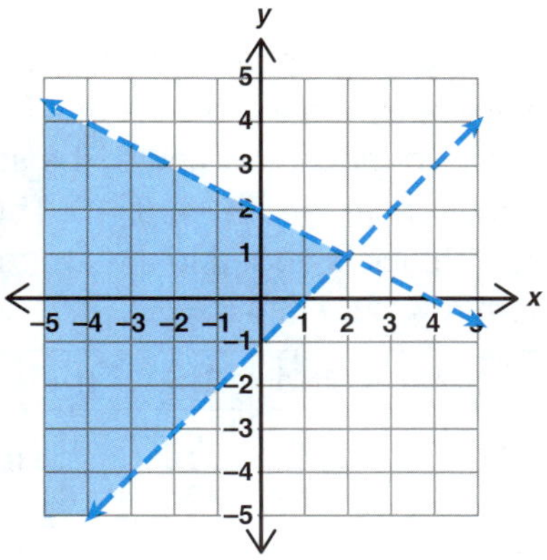

6.

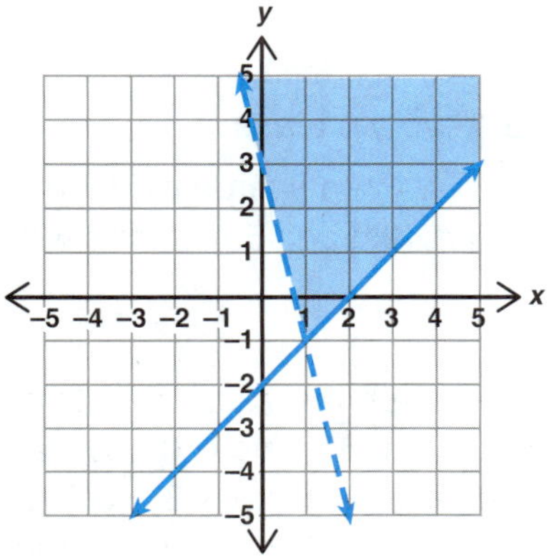

5

7.

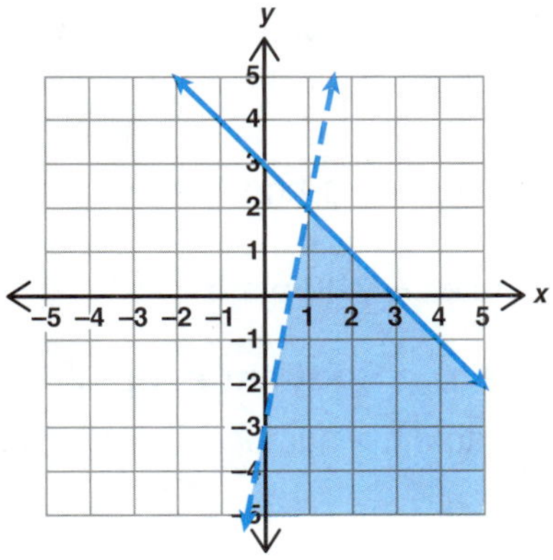

8.

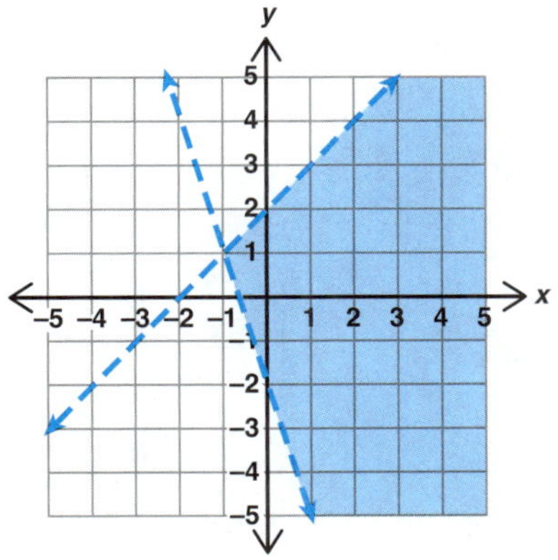

Interpret Graphs of Systems, pp. 88–89

1. **Job A**
   John would earn $60 for 5 sales because the job pays $60 per day no matter how many sales are made.
2. **14 sales**
   The line for Job C is at a higher point than the lines for jobs A and B at 14 sales.
3. **Job B**
   Job B is the best job if John can make from 7 to 12 sales per day. If he can't make that many, he should pursue Job A.
4. $40 \le x \le 60$
5. $x + y \le 90$
6. 60 T-shirts and 30 sweatshirts lead to the greatest profit.

   Test the individual points using the equation $P = \$15x + \$8y$.
   A: (40, 40) = $920
   B: (50, 40) = $1,070
   C: (60, 30) = $1,140
   D: (60, 25) = $1,100
   E: (40, 25) = $800

Unit 3 Review, pp. 90–91

1. $x = 3$, $y = -2$
2. **C. There is no solution.**
   The equations in the system have the same slope, so the graphs of the lines are parallel. Since the lines will never intersect, there is no solution.
3. **D. (1, –5)**
   Use substitution. Rewrite the first equation to solve for y: $-2x + y = -7$ becomes $y = 2x - 7$.

   Substitute the expression for y into the second equation and solve for x.

   $$-4x - 3y = 11$$
   $$-4x - 3(2x - 7) = 11$$
   $$-4x - 6x + 21 = 11$$
   $$-10x = -10$$
   $$x = 1$$

   Substitute the value for x into the first equation and solve for y.

   $$-2x + y = -7$$
   $$-2(1) + y = -7$$
   $$-2 + y = -7$$
   $$y = -5$$
4. **(0, –2)**
   Solve by elimination. To eliminate the x-variable, multiply every term in the first equation by 7 and every term in the second equation by 4. Then add. To eliminate the y-variable, multiply the first equation by 8 and the second equation by 3 before adding.
5. There are an infinite number of solutions to this system. If you were to graph both lines, they would lie on exactly the same points. Every point that is a solution to the first equation is also a solution to the second equation.

6.

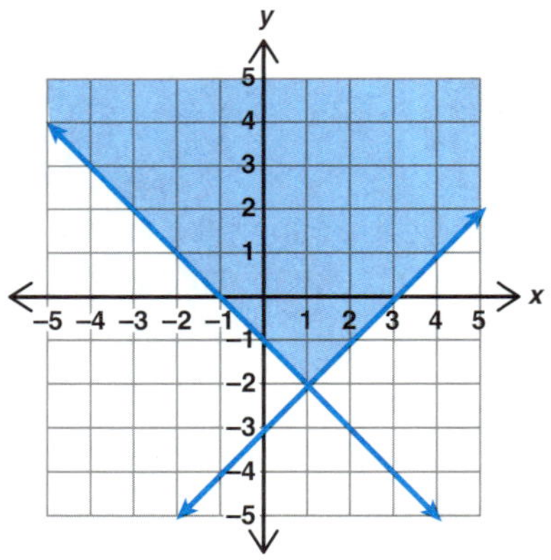

5

7. Yes

8. **17**

Let x = the 5-shelf units and y = the 4-shelf units. Write and solve the system.

$$\begin{cases} x + y = 32 \\ 5x + 4y = 145 \end{cases}$$

9. C. 4

10. **B. (–2, 3)**

Try the points in both inequalities.

11. ***Note:*** The order of the inequalities in the system doesn't matter.

$$\begin{cases} \mathbf{2L + 2W \leq 120} \\ \mathbf{L < 2W} \end{cases}$$

The fencing surrounds the garden, so you can apply the formula for finding the perimeter of a rectangle: $P = 2L + 2W$.

12. **C. 26**

Let x = the number of jeans sold and y = the number of shirts sold. Then write a system:

$$\begin{cases} 25x + 22y = 686 \\ 20x + 20y = 580 \end{cases}$$

Solve the system: x = 16 and y = 13.
The number of shirts sold is 13 × 2 days = 26 shirts.

UNIT 4

Patterns and Sequences, pp. 94–95

1. **An octagon (shape with 8 sides)**

2. **The letter J, blue**

J

3. **The letter U**

A is the 1st letter of the alphabet, E is the 5th, I is the 9th, M is the 13th, and Q is the 17th. The position of each letter is four more than the one before. The answer must be the 21st letter: U.

4. **55**

Each term is 9 less than the term before.

5. **37 blocks**

Your explanations may vary. Think of the top and bottom rows as separate from the blocks that form a square in the center. For each new figure, the square of blocks in the center increases: 1, 4, 9, 16. The next would be 25. Also, 1 block is added to the top row and 1 to the bottom row. 25 + 6 + 6 = 37

Terms	1st	2nd	3rd	4th	5th
Top and Bottom	4	6	8	10	12
Center Square	1	4	9	16	25
Total in Figure	5	10	17	26	**37**

6. ⊗

The pattern repeats every 6 terms:
⊗⊗⊠○○□. The open square will land on 6, 12, 18, . . ., 48. The 49th and 50th terms will be circles with Xs.

7. **3**

Use a calculator to convert the fraction $\frac{7}{11}$ to a repeating decimal: $\frac{7}{11} = 0.\overline{63}$. The first decimal place is 6 and the second decimal place is 3. Each odd numbered position is a 6, and each even numbered position is a 3. The 100th is an even position, so the decimal digit must be 3.

8. **2:40 p.m.**

The times move forward by 1 hour 15 minutes, or 75 minutes.
1:25 p.m. + 75 minutes = 2:40 p.m.

Finding the *n*th term in a Sequence, pp. 96–97

For problems 1–4, substitute the numbers 1 through 5 for n.

1. –4, 1, 6, 11, 16
2. –60, –54, –48, –42, –36
3. 13, 22, 31, 40, 49
4. –6, –7, –8, –9, –10
5. $-3n + 1$
6. $2n + 11$
7. $-10n + 7$
8. $12n - 19$
9. **–42**

The rule is $-4n - 2$, so the 10th term is $-4(10) - 2 = -42$.

10. **327**

The rule is $13n + 2$, so the 25th term is $13(25) + 2 = 327$.

11. **265**

The rule is $7n - 15$, so the 40th term is $7(40) - 15 = 265$.

12. **–2,375**
The rule is $-25n + 125$, so the 100th term is $-25(100) + 125 = -2{,}375$.

13. Your answers may vary. A sample answer follows: Doug's thinking uses a proportion to find the 30th term, but a sequence isn't a proportion. The rule $6n + 2$ creates this sequence, so the 30th term is $6(30) + 2 = 182$.

Working with a Geometric Sequence, pp. 98–99

1. **160**
Multiply by 2.
2. **16**
Divide by 2 or multiply by $\frac{1}{2}$.
3. **–243**
Multiply by –3.
4. **0.001**
Divide by 10 or multiply by $\frac{1}{10}$.
5. arithmetic
6. geometric
7. geometric
8. arithmetic
9. **–64**
Multiply by 2.
10. **8**
Divide by 5 or multiply by $\frac{1}{5}$.
11. **1,458**
$2 \times 3^6 = 1{,}458$
12. **1,000,000**
$\frac{1}{10} \times 10^7 = 1{,}000{,}000$
13. **38.4**
$0.15 \times 2^8 = 38.4$
14. **4,096**
$1 \times 4^6 = 4{,}096$

Function Basics, pp. 100–101

1. **a.** –13
b. 43
2. **a.** –7
b. 9
3. **a.** 3
b. 207
4. **a.** 23
b. 3
5. **a.** 11
b. 9
6. **a.** 1
b. 1
7. $f(n-2) = -2(n-2) + 1$
$= -2n + 4 + 1$
$= -2n + 5$
8. $g(-2a) = 3(-2a) - 2$
$= -6a - 2$
9. $h(\frac{b}{4}) = 4(\frac{b}{4}) - 3$
$= b - 3$
10. $\boldsymbol{x = 4}$
Substitute for n.
$f(n) = 4n - 5$
$f(x-1) = 4(x-1) - 5$
$= 4x - 4 - 5$
$= 4x - 9$
Set the expression equal to 7 and solve for x.
$4x - 9 = 7$
$4x = 16$
$x = 4$

Reading Graphs of Functions, pp. 102–103

1. **C.**
If you draw a vertical line between 0 and –1 on the x-axis, the line will cross the graph more than once.
2. –2
3. –1
4. –4
There is only one point on the graph where $f(x)$, or y, is equal to 4: (–4, 4).
5. **B.** $\boldsymbol{-4 \le x \le -1}$
At $f(-4)$, x is equal to 4. At $f(-1)$, x is equal to –5. The two points are 9 units apart.
6. Your answer may contain some of these ideas: Katie is correct. A rate compares two numbers. When Katie refers to rate of change, she means the slope of the interval. In the interval $4 \le x \le 5$, the ratio of the rise to run is 4 to 1. In each of the other intervals, the change is more gradual.

Finding Domain and Range of a Function, pp. 104–105

1. all real numbers, $\mathbb{R}$
all real numbers greater than or equal to 0, $x \ge 0$
all real numbers except for –5, $x \ne -5$
2. all real numbers, $\mathbb{R}$
all real numbers except for 0, $x \ne 0$
all real numbers greater than or equal to 4, $x \ge 4$
3. all real numbers except for 0, $x \ne 0$
all real numbers, $\mathbb{R}$
all real numbers greater than or equal to 0, $x \ge 0$
4. all real numbers greater than or equal to 0
all real numbers greater than or equal to 0
all real numbers, $\mathbb{R}$
5. all real numbers greater than or equal to 0
all real numbers, $\mathbb{R}$
all real numbers, $\mathbb{R}$

6. Domain: all real numbers, $\mathbb{R}$
   Range: all real numbers greater than or equal to –3
7. Domain: $x = -3$
   Range: all real numbers, $\mathbb{R}$
8. Domain: all real numbers, $\mathbb{R}$
   Range: all real numbers less than or equal to 3
9. Domain: all real numbers, $\mathbb{R}$
   Range: all real numbers greater than or equal to 3 or less than –1

Comparing Functions, pp. 106–107

1. **Alex**
   He took the least time to reach the 3-mile mark.
2. **Ron**
3. **Tom**
   The slope remained the same the entire race.
4. **Ron**
   His line never reaches the 10-mile mark. He stopped at 9 miles.
5. **Alex**
   He reached the 5-mile mark in the least amount of time.
6. Your answer should contain these ideas:
   Alex ran faster than the other runners for 3 miles. Then he rested for about 10 minutes. He ran the next 3 miles faster than the other runners, and then rested for almost 25 minutes. He ran the next 3 miles at the same speed as before. Then he slowed down for the final mile.
7. **About 300 pounds**
   The table shows an increase of about 50 pounds for every 2 inches in diameter.
8. **Blue spruce**
   Based on the table, the blue spruce saves 1,142 gallons of water per year.

   Use the function for the sugar maple:

   $g(x) = 4x^2 + 104x - 123$
   $= 4(8^2) + 104(8) - 123$
   $= 965$ gallons

   The blue spruce saves more water.
9. The sugar maple is much better than the apple tree on both measures. The blue spruce is about the same in reducing CO_2, but the blue spruce is much better than the apple tree at storm water recovery.

   Apple tree: 100 pounds CO_2
   150 gallons storm water

   Blue spruce: 106 pounds CO_2
   723 gallons storm water

   Sugar maple: about 160 pounds CO_2
   645 gallons storm water
10. **The blue spruce will be better because it grows more quickly.**
    The blue spruce grows 0.5 inch per year in diameter. In 10 years, it will grow 5 inches, giving it a 7-inch diameter. Use the table for storm water recovery. The number of gallons for a 7-inch-diameter tree should be halfway between 723 and 1,142—about 930 gallons.

    The sugar maple grows 0.3 inch per year in diameter. In 10 years, it will grow 3 inches, giving it a 5-inch diameter. Use the function for the sugar maple:

    $g(x) = 4x^2 + 104x - 123$
    $= 4(5^2) + 104(5) - 123$
    $= 497$ gallons

    After 10, years, the blue spruce will recover about 400 gallons more storm water annually than the sugar maple.

Functions in the Real World, pp. 108–109

1. 212°F
2. 32°F
3. Substitute –40° into the function.
   $f(x) = 1.8x + 32$
   $f(15) = 1.8 \cdot -40 + 32$
   $= -72 + 32$
   $= -40$
   The input is the same as the output. The temperature scales are equal at –40°.
4. **a.** 0.438 second
   b. 0.542 second
5. The snowy tree cricket chirps faster.
   Your reasoning process may vary, but your conclusion should be the same.
   Suppose a black cricket chirps 60 times per minute. The temperature would be 55°.
   How many times would a snowy cricket need to chirp in 13 seconds to be at a temperature of 55°?
   $x + 40 = 55$, so $x = 15$
   Compare the results. The black cricket is chirping 60 times per minute, or 15 times in 15 seconds. At the same temperature, the snowy cricket chirps 15 times in 13 seconds. The snowy cricket chirps faster.
6. **a. 56.9° F**
   Substitute 320 for x in the function.

   b. If x is 0, the temperature is 50.5° F. If x is 200, the temperature is 54.5° F. An increase of 200 parts per million causes an **increase of 4° F**.

   c. At 400 parts per million, the temperature is 58.5° F. If we reduce to 300 parts per million, the temperature would fall to 56.5° F—a decrease of 2° F.

Unit 4 Review, pp. 110–111

1. **circle**
   The pattern repeats in groups of 3. The 48th shape is a multiple of 3, so it is a square. Then the pattern starts again. The 49th is a circle and the 50th is a circle.
2. **A. 109**
   The pattern adds 13. By the tenth number, it will have increased 9 times. 13 × 9 = 117. Add this number to the first term: 117 + –8 = 109.
3. **B. 9*n* + 2**
4. **D. 3, 6, 12, 24, 48, . . .**
   The pattern is formed by multiplying instead of adding.
5. 1,875
6. –19
7. –11
8. –43
9. 60
10. 60
11. C. All real numbers except –5
12. C. –1
    When *y* or *f*(*x*) = 3, *x* = –1.
13. B. $-1 \le x \le 3$
14. D. All real numbers less than or equal to 0
15. No
16. Yes
17. **below average**
    Average height would be 40.1 inches.
18. Matt is at average height at 12 and 18 months. Then he has a growth spurt. He is above average at 24, 30, and 36 months.

Age (mo)	12	18	24	30	36
Matt's Height (in)	30	32	35	38	39
Average Height	30	32.3	34.2	35.9	37.4

UNIT 5

Simplify Polynomial Expressions, pp. 114–115

1. monomial
2. trinomial
3. monomial
4. trinomial
5. binomial
6. binomial
7. trinomial
8. monomial
9. binomial
10. monomial
11. $-4r^3 - 4 - 7r^3 + 3$
    $= -4r^3 - 7r^3 - 4 + 3$
    $\mathbf{= -11r^3 - 1}$
12. $7x - 1 - 6x^2 + 8$
    $= -6x^2 + 7x - 1 + 8$
    $\mathbf{= -6x^2 + 7x + 7}$
13. $4k + 6 - 6k + 2k$
    $= 4k - 6k + 2k + 6$
    $\mathbf{= 6}$
14. $-1 - 2y^2 + y^3 + 4 + 3y^2$
    $= y^3 - 2y^2 + 3y^2 - 1 + 4$
    $\mathbf{= y^3 + y^2 + 3}$
15. $-6n - 7n - 8 + 6n^2 + 4n$
    $= 6n^2 - 6n - 7n + 4n - 8$
    $\mathbf{= 6n^2 - 9n - 8}$
16. $2ab^2 - 6b - 8 + 3ab^2 - 2ab$
    $= 2ab^2 + 3ab^2 - 2ab - 6b - 8$
    $\mathbf{= 5ab^2 - 2ab - 6b - 8}$
17. $(4m - 2) + (-8m + 7m^2)$
    $= 4m - 2 - 8m + 7m^2$
    $= 7m^2 + 4m - 8m - 2$
    $\mathbf{= 7m^2 - 4m - 2}$
18. $(-3 - 4n) - (-5 + 6n)$
    $= -3 - 4n + 5 - 6n$
    $= -4n - 6n - 3 + 5$
    $\mathbf{= -10n + 2}$
19. $(-p^2 - p) + (7 + 5p^2)$
    $= -p^2 - p + 7 + 5p^2$
    $= -p^2 + 5p^2 - p + 7$
    $\mathbf{=4p^2 - p + 7}$
20. $(8x^3 + 8x^2 + 1) - (-5x^3 - 2x^2)$
    $= 8x^3 + 8x^2 + 1 + 5x^3 + 2x^2$
    $= 8x^3 + 5x^3 + 8x^2 + 2x^2 + 1$
    $\mathbf{= 13x^3 + 10x^2 + 1}$
21. $(3a - 8a^2 + 6) - (2a^2 + 6a + 7)$
    $= 3a - 8a^2 + 6 - 2a^2 - 6a - 7$
    $= -8a^2 - 2a^2 + 3a - 6a + 6 - 7$
    $\mathbf{= -10a^2 - 3a - 1}$
22. $(-2x - 8xy^2) - (6xy^2 - 2y - 7x)$
    $= -2x - 8xy^2 - 6xy^2 + 2y + 7x$
    $= - 8xy^2 - 6xy^2 - 2x + 7x + 2y$
    $\mathbf{= -14xy^2 + 5x + 2y}$

Multiply an Expression by a Monomial, pp. 116–117

1.	$2x^3$	$32ab^2$	$-10m^3n^4$
2.	$18n^2 + 24n$	$-21b + 6$	$10xy + 40y$
3.	$-64v^3 - 16v^2$	$-28m^3 - 7m^2$	$3x^3 - 4x^2$
4.	$km - 5km^2$	$30x^2y + 5xy^2$	$4a^4b + 32a^3b^2$
5.	$12p^2 - 3p + 15$	$-3n^2 + 4n + 3$	$16x^3 + 16x^2 - 40x$
6.	$-a^2 + 6a + 5$	$18v^3 + 36v^2 + 24v$	$x^3y + x^2y^2 - xy^4$

Core Connections: Modeling with Polynomial Functions, p. 117

1. **a.** $f(50) = 2.2(50) + \frac{1}{20}(50)^2 = 235$ ft

b. $f(75) = 2.2(75) + \frac{1}{20}(75)^2 = 446.25$ ft

2. $C = 500(30 - 0.02 \cdot 500) + 200$

$C =$ **\$10,200**

Multiply Binomials Using FOIL, pp. 118–119

1.	$b^2 + 6b - 7$	$r^2 + 6r - 16$	$k^2 + 6k + 5$
2.	$m^2 - m - 2$	$x^2 - 9x + 18$	$p^2 + 10p + 24$
3.	$n^2 + 19n + 84$	$b^2 - 6b - 40$	$3a^2 - 18a - 21$
4.	$7x^2 - 33x - 10$	$4n^2 + 24n + 27$	$10p^2 + 23p - 5$
5.	$6v^2 - 19v - 20$	$-9a^2 + 17a - 8$	$m^2 + 8m + 16$
6.	$n^2 - 25$	$4x^2 + 12x + 9$	$16k^2 - 1$

7. **D.** $\mathbf{x^2 + 11x + 18}$

$(x + 9)(x + 2) = x^2 + 11x + 18$

8. $\frac{1}{2}(4x + 2)(x + 6)$

$= \frac{1}{2}(4x^2 + 26x + 12)$

$\mathbf{= 2x^2 + 13x + 6}$

9. $(x + 10)(x - 3)$

$\mathbf{= x^2 + 7x - 30}$

10. Answers may vary but should include the following ideas:

Steve multiplied the first, inner, and last terms. He did not multiply the outer terms.

First: $x \cdot x = x^2$

Outer: $x \cdot -8 = -8x$

Inner: $8 \cdot x = 8x$

Last: $8 \cdot -8 = -64$

$x^2 - 8x + 8x - 64$

$= x^2 - 64$

Factor Out a Monomial, pp. 120–121

1. $3(x^2 + 11)$

$5y(3y^2 - 7)$

$-4m(9m^2 - 4)$

2. $5(5n^2 + 8n + 9)$

$8(5 - 2k - k^2)$

$-7x(5x^2 - 6x - 8)$

3. $10n(n^2 + 3n - 1)$

$2x^2(4x^2 + 8x + 7)$

$-9v(7v^3 + 4v - 5)$

4. $5a(18b^2 - 7b - 3)$

$8x(3x - 7y - 2)$

$5m(2mn^2 - m + 6)$

5. $2ab(5a + 10b + 4)$

$-7xy(7x^2 - 3y - 3)$

$3(4 - 9st + 6t^2)$

6. $-y(y^2 + x^2 + x)$

$8(5m^2 + 8n^2 - 9n)$

$3u^2v(5u + 3v + 3)$

7. B. $18mnp$

8. **A.** $\mathbf{4ab + 11b + 3}$

Factor out the monomial $8a$ from the expression.

$32a^2b + 88ab + 24a = 8a(4ab + 11b + 3)$

The width of the rectangle is the first factor. The length of the rectangle is the second factor.

The length of the rectangle is $4ab + 11a + 3$.

9. C. $SA = 2\pi r(r + h)$

10. Answers may vary but should include some of the following ideas:

Ginnie's answer is incorrect. Because 13 is not the greatest common factor of the expression, she has not completely factored the expression.

The greatest common factor is $13xy$. The correct factorization is $13xy(5x + 3y - 2)$.

Use FOIL to Factor Trinomials, pp. 122–123

1. $(x + 3)(x + 5)$

$(x - 7)(x + 4)$

$(x - 10)(x - 2)$

2. $(x + 8)(x - 4)$

$(x + 9)(x + 5)$

$(x - 6)(x - 8)$

3. $(x - 3)(x - 1)$

$(x - 3)(x + 2)$

$(x + 12)(x + 5)$

4. $(x + 16)(x + 5)$

$(x - 8)(x + 7)$

$(x + 12)(x - 9)$

5. $(2x + 3)(x - 7)$

$(2x - 7)(x - 1)$

$(3x - 5)(x + 2)$

6. $(3x - 2)(x + 6)$

$(5x + 4)(x + 2)$

$(2x - 3)(2x - 9)$

7. $(x + 10)(x - 10)$
$(x - 2)(x - 2)$ *or* $(x - 2)^2$
$(2x + 3)(2x - 3)$

8. $(x - 9)(x - 9)$ *or* $(x - 9)^2$
$(x + 11)(x + 11)$ *or* $(x + 11)^2$
$(x + 1)(x - 1)$

Mixed Review, pp. 124–125

1. $5n^2 + 3$ $\quad$ $2b^2 - 9b$

2. $2x^2 - 10x + 7$ $\quad$ $-7r^2 + 7r - 3$

3. $-14p^2 + p$ $\quad$ $3n^2 - 10n + 6$

4. $4m^2 - 6$ $\quad$ $-y^2 - 5y - 7$

5. $2x^2 + x + 4$ $\quad$ $-5r^2 - 8r + 3$

6. $-5a^3 - 6a^2 - a - 1$ $\quad$ $12v^3 + 12v^2 - 4$

7. $-12u^2v^2$
$7x^3y$
$-10m^2n$

8. $10a^2 - 2a$
$54u^2 - 36u^3$
$7xy - 49y^2$

9. $16n^2 + 64n - 24$
$28a^2 - 32a + 4$
$6x^3 - 30x^2 - 24x$

10. $-3p^3 + 18p^2 + 3p$
$28x^2y - 16xy^3$
$-5a^2b^2 + 5ab^3$

11. $n^2 + n - 30$
$x^2 + 14x + 33$
$y^2 - 20y + 36$

12. $2x^2 - 9x + 9$
$3m^2 - 26m + 48$
$4p^2 - 27p - 7$

13. $b^2 - 64$
$k^2 + 14k + 49$
$4x^2 - 4x + 1$

14. $k(k + 6)$
$2m(m - 13)$
$5u(2v + 3u)$

15. $9(x + 3y - 7)$
$mn(m + n)$
$-12b^2(a + 4)$

16. $17k(2k^2m - 1)$
$4a(a^2 - 10a + 8)$
$-11m(3m^3 + 7m - 9)$

17. $4xy(2x + 7y - 5)$
$uv^2(v - u + 1)$
$10k(8k^2 + 5j - 10)$

18. $(x - 7)(x - 2)$
$(v + 7)(v - 1)$
$(p - 10)(p - 9)$

19. $(b - 7)(b + 3)$
$(x + 4)(x + 13)$
$(m - 12)(m + 5)$

20. $(2n + 5)(n + 3)$
$(3y + 1)(y - 2)$
$(5k + 2)(k + 7)$

21. $(b + 10)(b + 10)$ or $(b + 10)^2$
$(x + 11)(x - 11)$
$(v - 3)(v - 3)$ or $(v - 3)^2$

22. $(x - 5)(x - 5) = \mathbf{x^2 - 10x + 25}$

23. $(5x + 2)(2x - 9)$

24. $a^2 + (a^2 - 3) + (a + 7) = \mathbf{2a^2 + a + 4}$

25. The student combined the terms $2x^2y$ and $-7xy^2$. These terms are not like terms because the variables are not raised to the same powers, so these terms should not be combined.
The correct answer is $2x^2y - 7xy^2 + 13y$.

Add and Subtract Rational Expressions, pp. 126–127

1. $\frac{2a - 3b}{10ab}$ $\quad$ $\frac{4m + 2n}{8n}$ $\quad$ $\frac{2u + v}{24u^3}$

2. $\frac{18mn + 20m}{12n}$ $\quad$ $\frac{-11u}{10}$ $\quad$ $\frac{24x^2 - 25x}{30x^2}$

3. $\frac{4x^2y + 24xy^2 - 9}{12xy}$ $\quad$ $\frac{5n^2 + 2m^2 - 16mn}{6mn}$ $\quad$ $\frac{32b + 9a^2 + 3ab}{12ab^2}$

4. $\frac{13}{12x^3 - 12x^2}$ $\quad$ $\frac{6 - 2m}{3m - 15}$ $\quad$ $\frac{2r + 7}{r^2 + 9r + 20}$

5. $\frac{4p^2 + 13p - 6}{(p + 2)(p + 4)}$ $\quad$ $\frac{-3k^2 + 40k}{5(k - 5)}$ $\quad$ $\frac{-12x^2 + 23x + 12}{(3x - 5)(x + 4)}$

6. $\frac{-n^2 + n - 6}{4(n + 1)}$ $\quad$ $\frac{2a^2 + a - 35}{(a - 1)(a + 7)}$ $\quad$ $\frac{n^2 + 12n + 24}{(n + 3)(n + 6)}$

7. $\mathbf{\frac{1}{x^2}}$

You can change a negative exponent to a positive exponent by moving the variable and its exponent to the opposite side of the fraction bar. x^{-2} is the same as $\frac{1}{x^2}$.

8. You can write $3y$ as a rational expression by writing it as a fraction over 1: $\frac{3y}{1}$.
Add $\frac{1}{2y} + \frac{3y}{1}$.

$\frac{1}{2y} + \frac{3y(2y)}{2y}$
$= \frac{1}{2y} + \frac{6y^2}{2y}$
$= \frac{6y^2 + 1}{2y}$

Multiply Rational Expressions, pp. 128–129

1.	$\frac{15v}{4}$	$\frac{27}{7mn}$	$\frac{4x}{5y}$
2.	$\frac{x+9}{3}$	$\frac{n+5}{n-1}$	$\frac{7p}{9}$
3.	$\frac{n+7}{n-8}$	$\frac{y+6}{9}$	$\frac{5}{2p}$
4.	$\frac{v-3}{13v}$	$\frac{1}{2m-7}$	$\frac{2x^2-5x-25}{(x-10)(x-6)}$
5.	$\frac{1}{n-3}$	$15x^2+45x$	1

Core Connections: Rational Expressions and Domain, p. 129

1. all real numbers except for –3, $x \neq -3$
2. all real numbers except for 0, $x \neq 0$
3. all real numbers except for $-\frac{3}{2}$, $x \neq -\frac{3}{2}$
4. all real numbers except for –5 and 4, $x \neq -5$ and $x \neq 4$

Divide Rational Expressions, pp. 130–131

1.	$\frac{5}{14x}$	$\frac{9km}{20}$	$\frac{a^3}{6}$
2.	$\frac{2a^2}{a+6}$	$\frac{x+6}{10x^2}$	$\frac{m-1}{7m}$
3.	$\frac{3}{2k}$	$\frac{r^2+4r-32}{4r}$	$\frac{7}{n-8}$
4.	$\frac{y}{(y+3)(y-12)}$	$\frac{m^2+6m+5}{4m^2(m-10)}$	$\frac{v^2-6v-16}{(v-14)(v+3)}$

Core Connections: Complex Fractions, p. 131

1. $14 \div \frac{2}{3} = \frac{14}{1} \cdot \frac{3}{2} = \mathbf{21}$
2. $\frac{4}{5} \div \frac{6}{35} = \frac{4}{5} \cdot \frac{35}{6} = \mathbf{\frac{14}{3}}$
3. $\frac{1}{1+\frac{1}{2}} = \frac{1}{\frac{3}{2}} = 1 \div \frac{3}{2} = \frac{1}{1} \cdot \frac{2}{3} = \mathbf{\frac{2}{3}}$
4. $\frac{2+\frac{1}{3}}{3-\frac{1}{3}} = \frac{\frac{7}{3}}{\frac{8}{3}} = \frac{7}{3} \div \frac{8}{3} = \frac{7}{3} \cdot \frac{3}{8} = \mathbf{\frac{7}{8}}$

Simplify Rational Expressions, pp. 132–133

1.	$\frac{x+1}{8}$	$\frac{n-8}{n-5}$	$\frac{a+2}{a-7}$
2.	$\frac{k+1}{10k}$	$\frac{5x^2}{2x-5}$	$\frac{1}{2(r-5)}$ or $\frac{1}{2r-10}$
3.	$x \neq 5, 11$	$y \neq 0, 7$	$n \neq -6, 2$
	$\frac{1}{x-11}$	$\frac{y-3}{y}$	$\frac{n-5}{n-2}$
4.	$n \neq -5, 5$	$x \neq -9$	$m \neq -8, -\frac{1}{2}$
	$\frac{n+10}{n+5}$	$\frac{3x}{x+9}$	$\frac{m-2}{2m+1}$
5.	$\frac{3^2+1}{3^3} = \mathbf{\frac{10}{27}}$		

6. $\frac{2(n-2)}{(2n+7)(n-2)} = \mathbf{\frac{2}{2n+7}}$

   $\frac{2}{2(-3)+7} = \frac{2}{1} = \mathbf{2}$
7. undefined
8. undefined
9. $\frac{3(x^2+4x+4)}{(x-9)(x+2)} = \frac{3(x+2)(x+2)}{(x-9)(x+2)} = \frac{3x+6}{x-9}$

   $\frac{3(1)+6}{1-9} = \mathbf{-\frac{9}{8}}$
10. $\frac{y-1}{(3y-8)(y+2)} \cdot \frac{(y-10)(y+2)}{(y-10)(y-1)} = \frac{1}{3y-8}$

    $\frac{1}{3(2)-8} = \mathbf{-\frac{1}{2}}$

Unit 5 Review, pp. 134–135

1. A. $x+6$
2. **D. 12**

   $\frac{(y+9)(y-2)}{(2y-5)(y-2)} = \frac{y+9}{2y-5}$

   $\frac{3+9}{2(3)-5} = \frac{12}{1} = 12$
3. C. m^2n^2
4. D. $\frac{v^2}{(v+2)(v+3)}$
5. D. $x = 10$
6. B. $8b^2$ and $3b^2$
7. binomial
8. monomial
9. trinomial
10. binomial
11. $5x(4x-9)$
12. $(x-13)(x+4)$
13. C. $\frac{4x^2+32x}{9x-63}$
14. $12mn$
15. D. $x+4$
16. A. $\frac{y^2+8y-6}{y^2+13y+30}$
17. $(m+1)(m+1)$ or $(m+1)^2$
18. **0**

    $\frac{r(3r-10)}{(3r-10)(r+2)} = \frac{r}{r+2}$

    $\frac{0}{0+2} = 0$
19. D. $\frac{u^3+uv+v^2}{u^2v}$
20. C. $x = 5, x = 7$
21. $\mathbf{2m^2 - mn - 7}$

22. Zoe is correct. First, factor the expression.

$$\frac{x^2 - 5x - 24}{x^2 + 18x + 45} = \frac{(x-8)(x+3)}{(x+15)(x+3)}$$

The expression is undefined at $x = -15$ and $x = -3$, so Sean is incorrect.

Next, simplify.

$$\frac{(x-8)(x+3)}{(x+15)(x+3)} = \frac{x-8}{x+15}$$

Finally, substitute 3 for x.

$$\frac{3-8}{3+15} = -\frac{5}{18}$$

This matches Zoe's answer.

UNIT 6

Quadratic Equation Basics, pp. 138–139

1.	{9, –9}	{0}	{10, –10}
2.	{8, –8}	$\{3\sqrt{2}, -3\sqrt{2}\}$ $\sqrt{18} = \sqrt{9 \cdot 2} = 3\sqrt{2}$	{2, –2}
3.	$\{\sqrt{10}, -\sqrt{10}\}$	$\{\frac{1}{3}, -\frac{1}{3}\}$	$\{\frac{3}{2}, -\frac{3}{2}\}$
4.	{4.8, –4.8}	{7.3, –7.3}	{9.7, –9.7}
5.	{3.2, –3.2}	{3.9, –3.9}	{3.3, –3.3}
6.	{1.7, –1.7}	{2.6, –2.6}	{7.7, –7.7}

Solve Quadratic Equations Using Graphs, pp. 140–141

1.

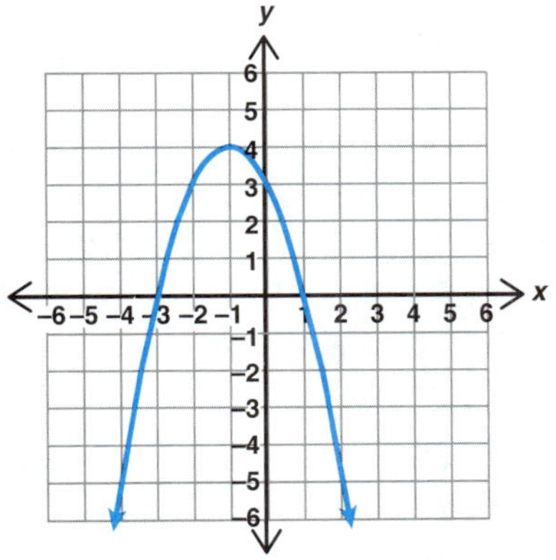

a. Your choice of points may be different.

x	*y*
–3	0
–2	3
–1	4
0	3
1	0

b. (–1, 4)

c. (–3, 0) and (1, 0)

d. domain: ℝ, all real numbers
range: All real numbers less than or equal to 4, $y \leq 4$

For problems 2–5, determine whether the graph of each equation opens up or down. Then substitute two points from the graph (roots, vertex) into the equation to see whether they make the equation true.

2. C

3. B

4. D

5. A

Solve Quadratic Equations by Factoring, pp. 142–143

Note: The order of the factors doesn't matter.

1. $(k+4)(k-4) = 0$
$\{-4, 4\}$
$n(n-8) = 0$
$\{0, 8\}$
$(m+6)(m-4) = 0$
$\{-6, 4\}$

2. $(x-8)(x-7) = 0$
$\{8, 7\}$
$(n-5)(n+4) = 0$
$\{5, -4\}$
$(a+4)(a-1) = 0$
$\{-4, 1\}$

3. $(w-8)(w+3) = 0$
$\{8, -3\}$
$(k-5)(k-7) = 0$
$\{5, 7\}$
$(b-4)(b-7) = 0$
$\{4, 7\}$

4. $(3x-1)(x+8) = 0$
$\{\frac{1}{3}, -8\}$
$(2n-1)(n-6) = 0$
$\{\frac{1}{2}, 6\}$
$(7m+1)(2m+1) = 0$
$\{-\frac{1}{7}, -\frac{1}{2}\}$

5. $(7n-6)(n-4) = 0$
$\{\frac{6}{7}, 4\}$
$a(5a-4) = 0$
$\{0, \frac{4}{5}\}$
$(2c-1)(c-3) = 0$
$\{\frac{1}{2}, 3\}$

6. C. $y = x^2 - 2x$

7. Your answers may vary. They may contain ideas similar to these:
Factoring is a way to find the roots, or solutions, of the equation. The roots tell where the graph intersects the x-axis.

Completing the Square, pp. 144–145

1. $n^2 - 14n + 49 = -25 + 49$
$$(n-7)^2 = 24$$
$$\sqrt{(n-7)^2} = \sqrt{24}$$
$$n - 7 = \pm 2\sqrt{6}$$
$\{7 + 2\sqrt{6}, 7 - 2\sqrt{6}\}$
2. $\{-3, -11\}$
3. $\{-5 + \sqrt{91}, -5 - \sqrt{91}\}$
4. $\{3 + \sqrt{37}, 3 - \sqrt{37}\}$
5. $\{-5 + \sqrt{94}, -5 - \sqrt{94}\}$
6. $\{17, -3\}$
7. $\{-4 + 3\sqrt{7}, -4 - 3\sqrt{7}\}$
8. $\{-9 + \sqrt{22}, -9 - \sqrt{22}\}$
9. $\{10, 8\}$
10. $\{-1.828, 3.828\}$

Use the Quadratic Formula, pp. 146–147

1. $\{3, -7\}$
Use $a = 1$, $b = 4$, $c = -21$
2. $\{\frac{5}{2}, -5\}$
Use $a = 2$, $b = 5$, $c = -25$
3. $\{2, -5\}$
Use $a = 1$, $b = 3$, $c = -10$
4. $\{\sqrt{2}, -\sqrt{2}\}$
Use $a = 1$, $b = 0$, $c = -2$
5. $\{2, \frac{2}{3}\}$
Use $a = 3$, $b = -8$, $c = 4$
6. $\{2 + \sqrt{5}, 2 - \sqrt{5}\}$
Use $a = 1$, $b = -4$, $c = -1$
7. $\{\frac{-1+\sqrt{13}}{4}, \frac{-1-\sqrt{13}}{4}\}$
Use $a = 4$, $b = 2$, $c = -3$
8. $\{3\sqrt{2}, -3\sqrt{2}\}$
Use $a = 1$, $b = 0$, $c = -18$
9. $\{\frac{5+\sqrt{37}}{3}, \frac{5-\sqrt{37}}{3}\}$
Use $a = 3$, $b = -10$, $c = -4$
10. $\{\frac{4+\sqrt{61}}{9}, \frac{4-\sqrt{61}}{9}\}$
Use $a = 9$, $b = -8$, $c = -5$
11. $\{1.3, -0.4\}$
12. Consider your answer correct if it includes these ideas:
The graph would open upward because the squared term is positive. The graph would cross the x-axis at (3, 0) and (–7, 0)—the roots of the equation.
Bonus idea: You may have been able to calculate the vertex by realizing that the x-value of the vertex must be exactly between 3 and –7, which equals –2. By substituting –2 for x in the equation, you get $(-2)^2 + 4(-2) - 21 = -25$. The vertex is at (–2, –25).

Solving Word Problems with Quadratic Equations, pp. 148–149

1. **14**
$x^2 + x = 210$
Discard the extraneous root: –15.
2. **2.37 seconds**
$-16t^2 + 90 = 0$
Discard the negative result.
3. **6 meters**
$$(x + 22)^2 = 784$$
$$x + 22 = \pm\sqrt{784}$$
$$x = -22 \pm \sqrt{784}$$
$$x = 6 \text{ or } -50$$
Discard the extraneous root: –50.
4. **80 meters by 130 meters**
$$w(w + 50) = 10{,}400$$
$$w^2 + 50w - 10{,}400 = 0$$
$$w = 80 \text{ or } -130$$
Discard the extraneous root: –130.
If the width is 80, the length is 80 + 50 = 130.
5. **25 inches**
$$(x + 9)^2 = x^2 + 369$$
$$x^2 + 18x + 81 = x^2 + 369$$
$$18x + 81 = 369$$
$$18x = 288$$
$$x = 16$$
If $x = 16$, then $x + 9 = 16 + 9 = 25$, the length of each side of the new square.
6. **20 seconds**
$$s = vt + \tfrac{1}{2}at^2$$
$$600 = 10t + \tfrac{1}{2}(2)t^2$$
$$600 = 10t + t^2$$
$$t^2 + 10t - 600 = 0$$
$$t = 20 \text{ or } -30$$
Discard the extraneous root: –30.

Unit 6 Review, pp. 150–151

1. **B. 8.4**
   $\sqrt{70} \approx 8.37$
2. **(–3, 4)**
   The vertex is the highest or lowest point on the graph.
3. **(–1, 0) and (–5, 0)**
4. **A. (–5, 1)**
   Find the points on the graph. See whether the points lie on the curved line.
5. **D. $y = 2x^2 - 8x + 5$**
   The point (0, 5) is on the line. When x is 0, then y is 5. Try this point in each equation. Only A or D could be correct. Then try another point, such as the vertex (2, –3). Only D is true for this point.
6. **B. $(3a - 8)(a + 1) = 0$**
   The solutions must make the binomial factors equal to 0. Try the solutions in the choices.
7. **{–2, 5}**
   You can use any solving method. This equation can be factored:
   $x^2 - 3x - 10 = 0$
   $(x - 5)(x + 2) = 0$
   $x = 5$ and $x = -2$
8. **C. 64**
   Take the b term –16. Divide by 2 and square the result. $-16 \div 2 = -8$ and $(-8)^2 = 64$
9. **D. 2.86**
   Plug in the information from the problem.
   $-16t^2 + 44t + 5 = 0$
   Use the quadratic formula and your calculator. The roots are {–0.11, 2.86}.
   Ignore the extraneous negative root. The ball will stay in the air for about 2.86 seconds.
10. **–6.5 or $-\frac{13}{2}$**
    This equation can be factored.
    $(2x + 13)(x - 3)$
11. **C. $1\frac{1}{2}$**
    Write an equation from the information in the problem.
    $(2x + 8)(2x + 10) = 63 + 80$
    $4x^2 + 36x + 80 = 143$
    $4x^2 + 36x - 63 = 0$
    Choose a method and solve.
12. **28**
    Write an equation from the information in the problem.
    $x^2 + (x + 17)^2 = 53^2$
    $x^2 + x^2 + 34x + 289 = 2{,}809$
    $2x^2 + 34x - 2{,}520 = 0$
    Choose a method and solve.
    One leg is 28 feet; the other leg is 45 feet.

Glossary

arithmetic sequence a pattern where each consecutive term increases or decreases by the same amount (p. 96)

bar graph a graph that uses bars to represent data amounts and compare categories (p. 20)

binomial an expression with two terms (p. 114)

$4x^2 + 5y$ $\quad$ $3y - 8$

box plot a visual way to show the minimum, lower quartile, median, upper quartile, and maximum of a set of data; also called a box-and-whisker plot (p. 38)

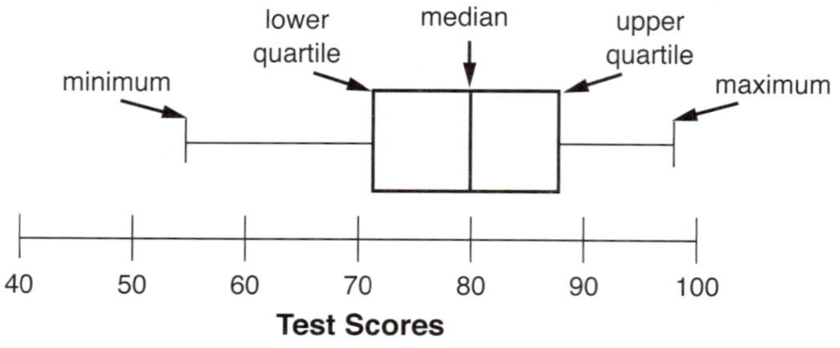

circle graph a graph that uses sections of a circle to represent data; a whole circle graph will equal 1, 100%, or $1 (p. 22)

combination an arrangement of items in a set in which the order doesn't matter (p. 62)

complex fraction a fraction within a fraction (p. 131)

$\frac{9}{\frac{1}{3}}$ is a complex fraction.

correlation the relationship between two sets of data (p. 24)

domain the first member in an ordered pair; the input (p. 100)

In the set {(3, 6), (4, 8), (5, 10)}, 3, 4, and 5 are part of the domain.

experimental probability the ratio of favorable outcomes to the number of trials (p. 70)

extraneous root a solution to a quadratic equation that does not make sense in the given situation (p. 148)

function a relation where each element of the domain results in exactly one element of the range (p. 100)

histogram a graph that uses bars to show frequency (p. 21)

like terms terms that have the same variable (p. 114)

$3x$ and $5x$ are like terms.
y^2 and $8y^2$ are like terms.

line graph a graph that displays data as points along a graphed line (p. 26)

line of best fit a straight line drawn through the core of plotted points on a scatter plot to help predict values (p. 44)

linear inequality an inequality that contains two variables (p. 86)

mean the average of a set of numbers, found by dividing the sum of the set by the number of values in the set (p. 28)

$5 + 7 + 12 = 24 \quad \frac{24}{3} = 8$

median the middle number of an odd set of values or the mean of the two middle numbers in an even set of values (p. 30)

5, 7, 12, 24, 30
↑
median

mode the value in a set of data that occurs most often (p. 30)

15, 7, 12, 24, 30, 24 The mode is 24.

monomial an expression with one term (p. 114)

$4x^2$ $5xy$ 88

normal distribution curve data represented by a curved line where most values fall beneath the center of the curve (p. 40)

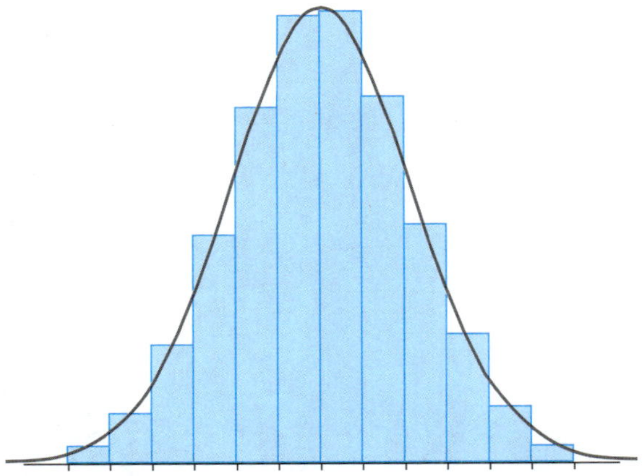

odds a ratio comparing the number of positive outcomes to the number of negative outcomes (p. 64)

outlier an extreme value in a set of data (p. 31)

parabola a symmetrical, curved line that is the graph of a quadratic equation (p. 140)

pattern a design or arrangement that has one or more repeating elements (p. 94)

permutation a counting strategy where order matters (p. 60)

polynomial an expression with one term or the sum (or difference) of unlike terms (p. 114)

population a large group targeted for data collection (p. 46)

probability the likelihood, or chance, of something happening or not happening (p. 64)

probability of an event = $\frac{\text{number of favorable outcomes}}{\text{total number of possible outcomes}}$

quadratic equation an equation with a variable raised to the 2nd power and no exponent greater than 2 (p. 138)

$y = 2x^2 - 7x + 4$ is a quadratic equation.

range the difference between the highest and lowest values in a set of data (p. 40)

range the second member in an ordered pair; the output (p. 100)

In the set {(3, 6), (4, 8), (5, 10)}, 6, 8, and 10 are part of the range.

roots the solutions to a quadratic equation (p. 140)

sample population a small segment of a larger population used for data collection (p. 46)

scatter plot a graph that uses data points to identify a possible correlation between groups of data (p. 24)

sequence a pattern that is defined by a rule (p. 94)

system of equations a collection of equations that use the same variables (p. 76)

term part of an expression: a variable, a number, or a variable multiplied by other variables or a number (p. 114)

trinomial an expression with three terms (p. 114)

$4x^2 + 9x + 5$ $\qquad$ $x + 5y + 3$

trinomial square a trinomial with identical binomial factors (p. 144)

$x^2 + 10x + 25 = (x + 5)^2$ $\qquad$ $x^2 - 8x + 16 = (x - 4)^2$

unlike terms terms that have different variables or variables raised to a different power (p. 114)

y and y^2 are unlike terms.
$8y^2$ and $8x^2$ are unlike terms.

vertex the lowest or highest point on a parabola (p. 140)

vertical line test a way to check whether a graph is a function. If a vertical line can pass through more than one point anywhere on a graph, the graph is not a function (p. 102).

weighted average an average where some values in the data set are worth more than others (p. 34)

zero property of multiplication when the product of two numbers is 0, at least one of the numbers must be 0 (p. 142).

If $ab = 0$, then either $a = 0$, $b = 0$, or both a and b equal 0.

Tool Kit

CALCULATOR BASICS FEATURING THE ONLINE CALCULATOR

A scientific calculator has many functions. You will need only a few of the functions to do the math on the GED® Math Test. Study the diagram to find the keys you will need. If you are using a different scientific calculator, the keys may be located in a different place, but they will have the same purpose.

Certain keys have more than one assigned function. To access a function written in small letters above a main key, first press the 2nd function key: [2nd]. Then press the key.

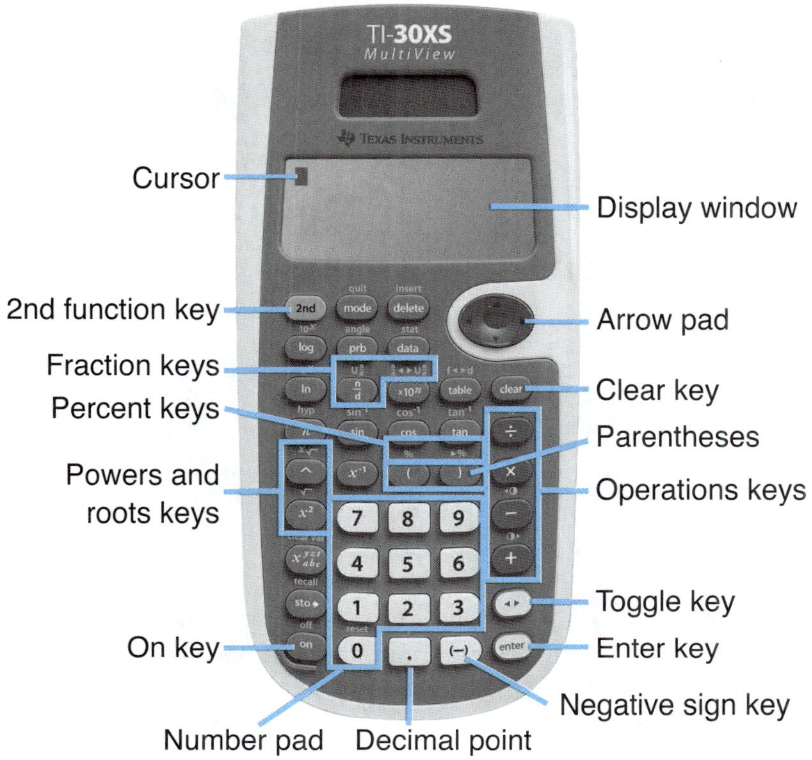

EXPLANATION OF SPECIAL KEYS

[clear]	Clears the display window. The memory is not erased.
[enter]	The enter key is used to process the operations you have entered. Think of it as an equal key (=).
x^2	Used to square a number.
^	Used to enter an exponent.
$\sqrt{\ }$ $\sqrt[x]{\ }$	Second functions used to find square roots and other roots.
( )	Used to enter parentheses or brackets to group operations.
[◄ ►]	The toggle key changes the form of the answer. It can change fractions or roots to decimals and back.
[(–)]	Used to enter a negative number. Do not enter a negative number using the subtraction symbol.
$\frac{n}{d}$	Used to input fractions.
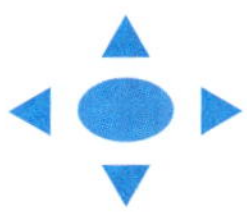	The arrow pad is used to move the cursor in the display window. Press an arrow to move in that direction.

COMPUTATIONS USING THE CALCULATOR

A scientific calculator always follows the order of operations. If you enter a chain of operations, it multiplies and divides before performing addition and subtraction steps. Use parentheses when a different order is needed. You can embed one set of parentheses inside another if more grouping symbols are needed.

Examples:	Press:	The display reads:	
$-5 \cdot 6 - 9$	[(–)] 5 × 6 – 9 [enter]	-5*6-9	-39
$-5(6 - 9)$	[(–)] 5 (6 – 9) [enter]	-5(6-9)	15
$\frac{-2+12}{-5}$	$\frac{n}{d}$ [(–)] 2 + 12 ▼ [(–)] 5 [enter]	$\frac{-2+12}{-5}$	-2
$6[-4 + 3(2 - 5)]$	6 ([(–)] 4 + 3 (2 – 5)) [enter]	6(-4+3(2-5))	-78

POWERS AND ROOTS

You can use your calculator to simplify and find roots and their decimal equivalents. You can also evaluate expressions containing exponents.

Examples:	Press:	The display reads:	
Simplify $\sqrt{48}$.	[2nd] $\sqrt{\ }$ 48 [enter]	$\sqrt{48}$	$4\sqrt{3}$
Find the decimal value of $\sqrt{48}$.	[2nd] $\sqrt{\ }$ 48 [enter] [◂▸]	$\sqrt{48}$ $4\sqrt{3}$ ◂▸	$4\sqrt{3}$ 6.92820323
Evaluate $\sqrt[3]{-64}$.	3 [2nd] $\sqrt[x]{\ }$ [(–)] 64 [enter]	$\sqrt[3]{-64}$	-4
15^2	15 [x^2] [enter]	15^2	225
$(-5)^4$	([(–)] 5) ^ 4 [enter]	$(-5)^4$	625
$4^5 - 3^7$	4 ^ 5 ▸ – 3 ^ 7 [enter]	$4^5 - 3^7$	-1163

FORMULAS

Perimeter (P)	Formula	Definition
Rectangle	$P = 2l + 2w$	l = length, w = width
Polygon with n sides	$P = s_1 + s_2 + \ldots + s_n$	s_1, etc. = each side
Circle	$C = \pi d$ or $C = 2\pi r$	C = Circumference, d = diameter, r = radius

Area (A)	Formula	Definition
Rectangle	$A = lw$ or $A = bh$	l = length, w = width, b = base, h = height
Square	$A = s^2$	s = side
Parallelogram	$A = bh$	b = base, h = height
Triangle	$A = \frac{1}{2}bh$	b = base, h = height
Trapezoid	$A = \frac{1}{2}h(b_1 + b_2)$	b_1 and b_2 = bases, h = height
Circle	$A = \pi r^2$	r = radius

Volume (V)	Formula	Definition
Rectangular/right prism (includes cube)	$V = Bh$	B = area of base, h = height
Cylinder	$V = \pi r^2 h$	r = radius, h = height
Pyramid	$V = \frac{1}{3}Bh$	B = area of base, h = height
Cone	$V = \frac{1}{3}\pi r^2 h$	r = radius, h = height
Sphere	$V = \frac{4}{3}\pi r^3$	r = radius

ADDITIONAL FORMULAS

Surface Area (*SA*)	Formula	Definition
Rectangular/right prism (includes cube)	$SA = ph + 2B$	p = perimeter of base, B = area of base, h = height
Cylinder	$SA = 2\pi rh + 2\pi r^2$	r = radius, h = height
Pyramid	$SA = \frac{1}{2}ps + B$	p = perimeter of base, s = slant height, B = area of base
Cone	$SA = \pi rs + \pi r^2$	r = radius, s = slant height
Sphere	$SA = 4\pi r^2$	r = radius

Other Useful Formulas	Formula	Definition
Simple interest (*I*)	$I = prt$	p = principal, r = rate, t = time
Distance (*d*)	$d = rt$	r = rate, t = time
Cost (*c*)	$c = nr$	n = number of items, r = rate per item
Slope of a line (*m*)	$m = \frac{y_2 - y_1}{x_2 - x_1}$	where (x_1, y_1) and (x_2, y_2) are points on the line
Slope-intercept form of the equation of a line	$y = mx + b$	m = slope, b = *y*-intercept
Point-slope form of the equation of a line	$y - y_1 = m(x - x_1)$	m = slope, (x_1, y_1) = a point on the line
Pythagorean theorem	$a^2 + b^2 = c^2$	a and b = legs of a right triangle, c = hypotenuse